PRECALCULUS: PATHWAYS TO CALCULUS
A PROBLEM SOLVING APPROACH
STUDENT WORKBOOK

MARILYN CARLSON – ARIZONA STATE UNIVERSITY

MICHAEL OEHRTMAN – UNIVERSITY OF NORTHERN COLORADO

KEVIN MOORE – UNIVERSITY OF GEORGIA

Fifth Edition

Rational Reasoning

Phoenix

HAYDEN
HM
McNEIL

Hayden-McNeil Sustainability

Hayden-McNeil's standard paper stock uses a minimum of 30% post-consumer waste. We offer higher % options by request, including a 100% recycled stock. Additionally, Hayden-McNeil Custom Digital provides authors with the opportunity to convert print products to a digital format. Hayden-McNeil is part of a larger sustainability initiative through Macmillan Higher Ed. Visit http://sustainability.macmillan.com to learn more.

Printed in the United States of America

10 9 8 7 6 5 4 3 2 1

ISBN 978-0-9963-8160-4

Hayden-McNeil Publishing
14903 Pilot Drive
Plymouth, MI 48170
www.hmpublishing.com

CarlsonM 8093-2 F15

Introduction

Overview of Workbook Content

This workbook contains investigations and homework for the text's eleven modules. The investigations introduce the modules' central ideas and include questions to help you build your own understanding of these ideas. Each module in this workbook has three to ten investigations and one homework set. The sections of the homework follow the naming and ordering of the worksheets and text. The text contains online videos that explain key ideas and explain examples like ones in your homework—we urge you to use these videos to help you complete your homework and strengthen your understandings. We have intentionally included problems in each section of the homework that are much like problems in the investigations and in the videos. This is so you can develop the thinking and understandings that will be needed to be successful in calculus.

The first module focuses on understanding and practicing methods of simplifying expressions, evaluating functions and solving equations. This module has three investigations that provide review of these important skills. You can also find many practice problems in Module 1 of the online text. All you need to do is just follow the links.

Modules 2 – 11 focus on building your understanding of standard precalculus ideas that are foundational for calculus. By engaging with the questions in the investigations and homework, your reasoning and problem-solving abilities will get better and better. Over time you will become a powerful mathematical thinker who has confidence in your ability to figure out novel problems on your own.

To: The Precalculus Student

Welcome!

You are about to begin a new mathematical journey that we hope will lead to your choosing to continue studying mathematics. Even if you don't currently view yourself as a math person, it is very likely that these materials and this course will change your perspective. The materials in this workbook are designed with student learning and success in mind and are based on decades of research on student learning. In addition to becoming more confident in your mathematical abilities, the reasoning patterns, problem solving abilities and content knowledge you acquire will make more advanced courses in mathematics, the sciences, engineering, nursing, and business more accessible. The investigations and text will help you see a purpose for learning and understanding the ideas of precalculus, while also helping you acquire critical knowledge and ways of thinking that you will need for learning calculus. To assure your success, we urge you to take advantage of the many resources we have provided to support your learning. We also ask that you make a strong effort to make sense of the questions and ideas that you encounter. This will assure that your mathematical journey through this course is rewarding and transformational.

Wishing you much success!

Dr. Marilyn P. Carlson, Dr. Michael Oehrtman, Dr. Kevin Moore

and members of our project team: Alan O'Bryan, Michael Tallman, Kristin Frank, Ben Whitmire, Grant Sander, Ashley Duncan, Caren Burgermeister, and Tim Persson

Table of Contents

This module develops foundational ideas of angle and angle-measure by investigating approaches for measuring the openness of two rays. We introduce methods for modeling the behavior of periodic motion in the context of co-varying an angle measure with a linear measurement that maps out a periodic motion, laying the groundwork for introducing the trigonometric functions of sine and cosine. The module concludes by exploring the meaning of period, amplitude, and translations of both the sine and cosine functions.

We explain the relationship between right triangle and unit circle trigonometry by initially exploring the right triangle relationships defined by sine, cosine, and tangent functions in a unit circle context. The triangle relationships defined by the sine, cosine, and tangent functions are used to determine the values of unknown quantities in various applied problems. We conclude the module by deriving various trigonometric identities that relate the trigonometric functions to one another.

We introduce polar coordinates with examples that illustrate the usefulness of polar coordinates as an alternate coordinate system. Students practice graphing polar functions and using polar functions to track the co-variation of distances and directions to plot and interpret the graphs of these functions. We next spend time exploring how constant rate of change is represented in polar coordinates and how to identify the graphs of functions with a constant rate of change. Similarly, we explore the meaning of the average rate of change for functions defined in polar coordinates. After working with average rates of change, we determine methods for converting between rectangular and polar coordinates and converting functions defined in one system to a function in the other system that produces the same graph.

This module begins by exploring what a vector is and how to represent a vector in both polar form and component (or rectangular) form. We next investigate vector addition as a way of determining the net effect of multiple vectors interacting with one another. We conclude the module by learning how to scale vectors (the process of changing the magnitude of a vector by some factor) with both positive and negative scalars.

We begin this module by introducing terminology and then learn how to graph sequences by generating ordered pairs consisting of the position and value of the terms. We introduce notation for describing sequences and how to define patterns using both recursive and explicit formulas. We apply these ideas by developing the meaning of arithmetic and geometric sequences and develop formulas to represent them. We then define what we mean by a series and use sequences of partial sums to keep a running total of the sum of the terms of the sequence. The module concludes by establishing sigma notation as a way of representing a series and by developing methods for finding the sums of finite arithmetic series, finite geometric series, and infinite geometric series.

Simplify or Expand to Produce Equivalent Expressions

Order of Operations

1. Evaluate the following:

   a. $-2 + 5 - 12 =$

   b. $6 - (-4) + 1 =$

   c. $\dfrac{-5 + (9 - 5(4)) + 3}{5(-2) - 1} =$

2. Evaluate the following:

   a. $\left| -5 - (-2) \right| =$

   b. $\left| \dfrac{-3(5-1)}{-4} \right| =$

3. Suppose that $a = 2$, $b = -3$, $c = 0$, and $d = -5$. Find the value of the following expressions.

   a. $b(2a - 3d)$

   b. $d^2 - 25 + c$

   c. $6 - \dfrac{b}{ad}$

   d. $-a\sqrt{b^2 + c^2}$

4. a. Simplify the following:

   i. $\dfrac{(2)(2)(2)(2)(2)}{(2)(2)(2)}$

   ii. $\dfrac{(7)(7)(7)}{(7)(7)(7)}$

   iii. $\dfrac{(5)(5)}{(5)(5)(5)(5)(5)}$

   iv. $\dfrac{(x-3)(x-3)}{(x-3)} =$

   v. $\dfrac{(x+2)^2(x-1)}{(x+2)^5(x-1)(x+1)} =$

   vi. $\dfrac{x^3 y^2 z^2}{x^5 y z^4} =$

   b. Explain why $\dfrac{x^m}{x^n} = x^{m-n}$.

c. Expand and multiply to justify why $\left(x^3\right)^5 = x^{15}$. Explain why $\left(x^m\right)^n = x^{(m)(n)}$.

d. Simplify: $\dfrac{96x^6 y^7 z^9}{128x^6 (y^3 z)^4} =$

e. Explain why $x^0 = 1$.

5. Simplify the following. Verify that your simplified expression is equivalent to the original expression by substituting several different values for x. If the expression does not simplify, write DNS.

   a. $2(x^2 + 1) =$

   b. $-2(x - 5) =$

   c. $-2x(5x^2 + 4) =$

   d. $\dfrac{-15x + 7}{3} =$

   e. $\dfrac{6x^4 - 5}{-2x} =$

   f. $\dfrac{-6x^5 + 7x}{-3x} =$

6. Are the following statements valid? Explain and verify by substituting numbers in for the variables and evaluating.

   a. $4y = 4x + 5$ implies that $y = x + 5$

   b. $\dfrac{ab + c}{a} = b + c$

   c. $-5a(-7a + 5) = 35a^2 - 25a$

   d. $\dfrac{2y - 6x}{3x} = 2y - 2$

7. Simplify: $\dfrac{1}{5}\left(30x^4 + 12x^2 - 3x - 5\right) - \left(2x^4 - 6x^2 + 5x\right)$

8. Multiply each of the following to obtain equivalent expressions.

    a. $(x-3)(x+9)=$

    b. $(2x-7)(-3x+2)=$

    c. $(x+5)^2 =$

    d. $(3m-2n)^2 =$

    e. $(3x-4y)(5z+8x^2)=$

9. Draw a diagram to illustrate why the area of a square with side length, $s = 4$ kilometers, is computed using the formula $A = s^2$. What is the unit of the answer? Explain.

10. Illustrate on the given diagram why the area of a square with side length $a+b$ units is equal to $a^2 + 2ab + b^2$ square units.

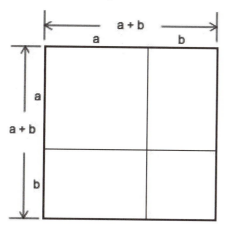

11. Is it true that $(a+b)^2 = a^2 + b^2$? Explain.

12. Factor each of the following to obtain equivalent expressions.

    a. $x^2 - 2xy + y^2 =$ b. $3a^2 + 6ab + 3b^2 =$

    c. $2x^3 + 5x^2 - 12x =$ d. $16x^2 - 36 =$

13. Simplify each of the following expressions. If the expression does not simplify, write DNS.

    a. $\dfrac{x^2 + 5x - 14}{x - 2}$ b. $\dfrac{x + 1}{3x^2 - x - 4}$

    c. $\dfrac{x^3}{x^3 - 2x^2 + 5x}$ d. $\dfrac{x^2 - 7x + 2}{x}$

14. Simplify the following radicals to obtain equivalent expressions. If the expression does not simplify, write DNS.

    a. $\sqrt{75x^9 y^2 z}$ b. $\sqrt{\dfrac{a^{11}}{b^4 c^7}}$ c. $\sqrt[3]{24a^5 b^2 c^3}$

    d. $\sqrt{a^2 - b^2}$ e. $\sqrt{a^2 + 9}$ f. $\sqrt{a^2 + 2ab + b^2}$

15. A student has simplified the following expressions **incorrectly**. Identify the student's error and then substitute numbers in for the variables to verify that the expressions are not equivalent. **Correct the error by writing an expression that is equivalent to the one on the left side of the equal sign.**

a.

$$(2a - 3b)^2 = 4a^2 - 9b^2$$

b.

$$3x(x-2) - 4(5-x) = 3x^2 - 6x - 20 - x$$
$$= 3x^2 - 7x - 20$$

c.

$$3(2x-5)^2 = (6x-15)^2$$
$$= 36x^2 - 180x + 225$$

d.

$$\frac{25x^3 - x}{5x} = 5x^2 - x$$

e.

$$\sqrt{x^2 + 16} = x + 4$$

f.

$$\sqrt{x^2 - 25} = x - 5$$

g.

$$\sqrt{-18} = -3\sqrt{2}$$

1. a. Indicate on the figure and describe the attribute of the rectangle below that represents the following quantities.

l = length of the side of the rectangle measured in inches
w = width of the side of the rectangle measured in inches
p = perimeter of the rectangle measured in inches
A = area of the rectangle measured in square inches

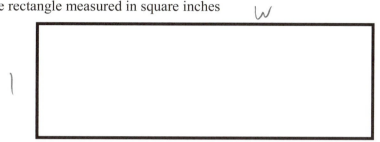

$P = 2L + 2w$
$A = L \cdot w$

b. Write the formula that determines the perimeter of a rectangle, given its width and length.

$P = 2L + 2w$

c. Determine the width w of a rectangle when $p = 21$ inches and $l = 7$ inches. Show your work.

$21 - 14 = \dfrac{7}{2}$

$\boxed{w = 3.5}$

d. Write the formula that determines the area of a rectangle, given its width and length.

$A = L \cdot w$

e. Determine the area A of a rectangle when $p = 26$ inches and $l = 5$ inches. Show your work.

$26 - 10$

$\dfrac{16}{2} = 8.5$

$w = \dfrac{8}{1} = 7 = 56$

2. a. Indicate on the figure and describe the attribute of the circle below that represents the following quantities.

$A = \pi \left(\dfrac{d}{2}\right)^2$ $A = \pi r^2$

$C = 2\pi r$

r = radius of the circle measured in feet
d = diameter of the circle measured in feet
C = circumference of the circle measured in feet
A = area of the circle measured in square feet

$r = \frac{1}{2}$ diamiter
$d =$ length of a line that crosses through the center of a circle.
$A =$ the space inside the circle
$C =$ the length of the line of the circle

b. Write the formula to express a circle's diameter in terms of its radius. Use the formula to determine the diameter of a circle that has a radius of 4.721 feet.

$4.721 \cdot 2 = 9.442$ ft

c. Write the formula to express a circle's circumference in terms of its diameter. Use the formula to determine the circumference of a circle that has a diameter of 6.48 feet.

$$\frac{6.48}{2} \qquad 3.24\,ft \qquad 2\pi(3.24) = 20.3472\,ft.$$

d. Write the formula to express a circle's area in terms of its diameter. Use the formula to determine the area of a circle that has a diameter of 4.09 feet.

$$\frac{4.09}{2} \qquad 2.045 \qquad \pi\left(\frac{4.09}{2}\right)^2 = 13.13\,ft.$$

e. Write the formula to express a circle's circumference in terms of its radius. Use the formula to determine the circumference of a circle that has a radius of 3.5 feet.

$$2\pi\,3.5 \qquad = 21.98\,ft$$

f. Write the formula to express a circle's area in terms of its diameter. Use the formula to determine the area of a circle that has a diameter of 3.5 feet.

$$A = \pi\left(\frac{d}{2}\right)^2$$

$$3.14\left(\frac{3.5}{2}\right)^2 = 9.616\,ft$$

g. Define a formula to express the circumference of a circle in terms of its area. Use the formula to determine the circumference of a circle that has an area of 42.7 square feet.

$$C = 2\sqrt{\pi A} \qquad 2\sqrt{\pi\,42.7}$$

3. a. Indicate on the figure and describe the attribute of the cube below that represents the following quantities.
   s = length of the sides of the cube measured in centimeters
   SA = surface area of the cube measured in square centimeters
   V = volume of the cube measured in cubic centimeters

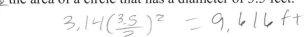

$$V = S^3$$

b. Write the formula to express the volume of a cube V in terms of its side length s. Determine the volume of a cube that has a side length of $4\frac{5}{8}$ centimeters.

$$4.625^3 = 98.93\,cm^3$$

c. Write the formula to express the surface area of a cube SA in terms of its side length s. Determine the surface area of a cube when the cube's side length is $2\frac{2}{3}$ centimeters.

$$2\frac{2}{3}^2 = 7.\overline{111} \cdot 6 = 42.\overline{666} \ cm^2$$

d. Define a formula to express the volume of a cube V in terms of its surface area SA. Determine the volume of a cube that has a surface area of 62 square centimeters.

$$\frac{62}{6} = 10.333 \qquad V = \left(\frac{SA}{6}\right)^{3/2} \qquad 33.217 \ cm^3$$

4. a. How much greater is the area of a square with a side length of 8 inches than the area of a circle with a radius of 3 inches?

b. Make a drawing to illustrate a circle with a radius of 3 inches inside a square with a side length of 8 inches and shade the area that represents the difference between these two areas.

c. When the area of a square with side length l is larger than the area of a circle with radius r, define a formula to express the difference between the two areas. What is the side length of the square, if the difference in area is 2 square inches and the radius of the circle is 3 inches?

1. A candle that is originally 10 inches long has burned x inches.
   a. Make a drawing of the 10-inch candle and illustrate the quantities 10 inches, $10 - x$ inches, and x inches on your drawing. What does $10 - x$ represent?

   $10 - x =$ the current length of the candle

   b. Define a formula to express the length of the candle y in terms of the number of inches that have burned from the candle x.

   $$y = 10 - x$$

   c. Determine the value of x when the length of the candle y is 3.4 inches. What does the solution to this equation represent?

   $x = 6.6$ how much the candle had removed

   $$y = 10 - x$$
   $$x = 10 - y$$
   $$x = 10 - 3.4 = 6.6$$

   d. Construct a graph that represents the length of the candle y in terms of the number of inches that have burned from the candle x. Label two points on your graph and describe what each point represents.

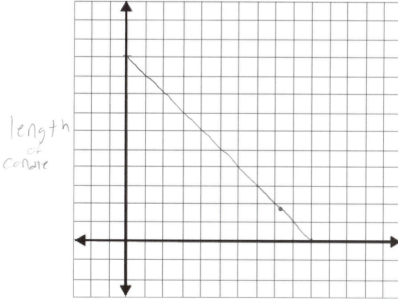

length of candle

amount burnt off

e. As the length of the candle decreases, how does the number of inches that have burned from the candle change?

for each in lost, the # of in. burt increses

f. Set up an equation to algebraically determine the number of inches that have burned from the candle when the length of the candle is 1.4 inches. Illustrate your solution on the graph in part d.

$x = 10 - y$
$x = 10 - 1.4$
$x = 8.6 in$

g. Describe what it means to "solve for x" in the context of this problem.

Find how many inches have burnt off.

2. Write an equation for each of these situations and then *solve the equation* for the unknown value. Define a variable to represent the value of the unknown number before writing each equation.
    a. 17.5 is equal to 2 times some number. What is the number?

$2x = 17.5$
$x = 8.75$

    b. The sum of 3 times some number and 12 is 42. What is the number?

$3x + 12 = 42$
$3x = 30$
$x = 10$

    c. $\frac{1}{4}$ of some number is 4.3. What is the number?

$\frac{.25x}{.25} = \frac{4.3}{.25}$
$x = 17.2$

    d. The difference between some number and 14.3 is 2.1. What is the number?

$x - 14.3 = 2.1$
$x = 16.4$

    e. Some number is 4 times as large as 9.8. What is the number?

$x = 9.8 \cdot 4$
$x = 39.2$

f. Some number is equal to $\frac{1}{3}$ of the sum of 88.2, 93.5 and 64. What is the number?

$$x = \frac{1}{3}(88.2 + 93.5 + 64) \qquad x = 81.9$$

g. 45 is some multiple of 15. Set up an equation and determine the value of the multiple.

$$15x = 45 \qquad x = 3$$

h. $1,200 is 1.5 times as large as some number. What is the number?

$$\frac{1200}{1.5} = \frac{1.5x}{1.5} \qquad x = \$800$$

i. 200% of some number is 38.2. What is the number?

$$2x = 38.2$$
$$x = 19.1$$

j. $\frac{3}{4}$ of some number increased by 12 is equal to 5 times the number. What is the number?

$$\frac{3}{4}x + 12 = 5x \qquad x = \frac{48}{17}$$
$$12 = 4.25x$$

3. The distance between some number(s) and 0 on the number line is 4.
   a. What are the numbers?

   b. Illustrate on the given number line all numbers that are 4 units away from 0.

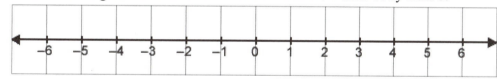

   c. Solve the equation $|x| = 4$ for x (what values of x make this equation true)?

   d. Describe how the solutions found in part (c) are represented on the number line.

4. Solve the following equations for the unknown value(s) (i.e., solve for x).
   a. Given that $|x| = 5$, determine the value(s) of x that makes this equation true.

b. Given that $2x - 19 = 5$, what value(s) of x makes this equation true.

c. Given that $\sqrt{x} = 4$, what value(s) of x makes this equation true.

d. Given that $x^2 = 36$, what value(s) of x makes this equation true.

e. Given that $\sqrt[3]{x} = -2$, what value(s) of x makes this equation true.

5. a. Given a graph that represents how x and y are related, determine the value of x when $y = 8$. Describe your approach and say why it works.

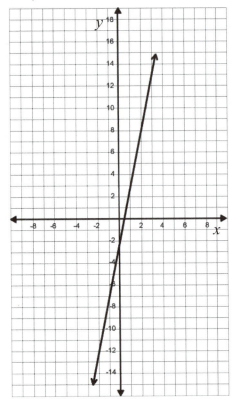

$y = 5x - 2$

b. Given that $y = 5x - 2$ is the equation of the line graphed above, solve the equation $8 = 5x - 2$ for x algebraically. Illustrate how to use the graph in part a to solve this equation.

c. Use the graph in part a to solve the equation $3 = 5x - 2$. Illustrate how you used the graph to solve this equation.

d. Use the graph in part a to solve the equation $0 = 5x - 2$. Illustrate how you used the graph to solve this equation.

6. Solve the following equations for x:

a. $\dfrac{3}{4}x + \dfrac{5}{6} = 5x - \dfrac{125}{3}$

b. $\sqrt{x+1} - 3x = 1$

c. $|5x - 6| + 3 = 10$

d. $4 = \dfrac{x+3}{x-1}$

7. A student has **incorrectly** solved the following equations for x when $y = 0$. Describe the mistake(s) made by the student. Then provide a **correct** solution for each of the equations.

a.
$$y = -5 + \sqrt{x^2 + 16} \quad ; \quad y = 0$$
$$\text{Solution:} \quad 0 = -5 + \sqrt{x^2 + 16}$$
$$0 = -5 + x + 4$$
$$0 = -1 + x$$
$$\boxed{1 = x}$$

b.
$$y = x^3 - 3x^2 - 4x \quad ; \quad y = 0$$
$$\text{Solution:} \quad 0 = x^3 - 3x^2 - 4x$$
$$\frac{0}{x} = \frac{x^3}{x} - \frac{3x^2}{x} - \frac{4x}{x}$$
$$0 = x^2 - 3x - 4$$
$$0 = (x + 4)(x - 1)$$
$$\text{Thus,} \quad x + 4 = 0 \qquad x - 1 = 0$$
$$\boxed{x = -4} \qquad \boxed{x = 1}$$

*0. Over the holidays, you and your friends drove from Phoenix to Flagstaff for a ski trip. While pulling out of your driveway you noticed that your car's speedometer was broken. Since you had received a speeding ticket the prior week, it was important that you kept track of your speed to avoid receiving another ticket.

a. What quantities could you use to **estimate** your speed as you pass by the Montezuma Castle exit (a popular speed trap) on your drive to Flagstaff?

b. What units could you use to measure each of these quantities?

c. Choose two specific values for each of the quantities described in (a) and (b) and explain how to use these values to estimate the speed of the car.

d. Is it possible to determine your exact speed as you pass by Montezuma's Castle? Does your speedometer report your instantaneous speed? Explain.

*1. Consider a chair in your classroom.

a. Identify five attributes of the chair that can be measured.

b. Which of the attributes you identified in Part (a) are fixed quantities? Explain.

c. Are any of the attributes you identified in Part (a) varying quantities? Explain

d. Identify three attributes of the chair that cannot be measured.

e. Imagine the chair moving from the east wall to the west wall of a classroom. Is the chair's distance (in feet) from the east wall a quantity? If so, is it fixed or varying? Is the chair's distance (in feet) to the west wall a quantity? If so, is it fixed or varying?

f. Suppose that the distance across the classroom, from the east wall to the west wall, is 64 feet. Draw a diagram that represents this distance. Does the distance between the two walls vary or is this distance constant/non-varying?

g. Suppose that the chair has wheels. Imagine that you push the chair so that it rolls straight across the room from the east wall to the west wall. Illustrate, in the diagram you created above, the location of the chair when it has moved about 1/3 of the way across the classroom from the east wall. (How might you mark your diagram so that it shows the chair continues to move until it hits the west wall?)

h. If you know the chair has moved 50 feet away from the east wall, how can you determine the chair's distance (in feet) to the west wall?

i. What two quantities did you use to compute the chair's distance (in feet) to the west wall? How did you combine or compare those two quantities (using addition, subtraction or multiplication) to determine your answer?

j. Represent the chair's distance (in feet) from the west wall "in terms of" the chair's distance from the east wall by using words to complete the statement on the right side of the equal sign. The words you use should reference the distance from the east wall.

the chair's distance (in feet) from the west wall = _____

The attributes of an object that you can imagine as being measured are called **quantities**. To clearly describe a quantity, we must mention the object that is being measured, the attribute of the object that is being measured, and the units used in the measurement.

If the value of a quantity does not change the quantity is called a **fixed quantity** and its value is a **constant.**

If the value of the quantity does change the quantity is called a **varying quantity** and the quantity can assume more than one value.

Examples of *fixed quantities*: the volume of a box (in cubic inches) and the weight of a table (in pounds).
Examples of *varying quantities*: the distance (in miles) of your car from home as you drive to work, the height (in feet) of a ball above the ground after it is thrown, the volume of water (in gallons) that has been added to an empty fish tank as it is being filled.

2a. Imagine an empty bathtub. You turn on the water and it begins to fill. Identify at least four quantities that are associated with this situation. You may identify both fixed and varying quantities. (Be specific in describing the attribute being measured, including where it is being measured from.)

Object: _____   Object: _____

    Attribute: _____   Attribute: _____

    Fixed or Varying?: _____   Fixed or Varying?: _____

    Choice of unit: _____   Choice of unit: _____

Object: _____   Object: _____

    Attribute: _____   Attribute: _____

    Fixed or Varying?: _____   Fixed or Varying?: _____

    Choice of unit: _____   Choice of unit: _____

  b. When defining a quantity why is it important to describe where the quantity is being measured from?

3a. Currently, the following statements do not properly define varying quantities. Rewrite each of the following statements so that they each appropriately define a specific varying quantity.

    i. gallons of gasoline

    ii. distance traveled in feet

    iii. height

  b. For each quantity you defined in part (a), illustrate with a number line what is being measured. (Draw a number line, place a vertical hash at 0, and label what 0 represents.)

    i.

    ii.

    iii.

In mathematics we want to represent how the values of two or more quantities are related and how their values co-vary or change together. Instead of representing a quantity's values with words it is more convenient to use a shorter notation. This is the purpose of ***defining variables*** to represent all the values that a specific varying quantity might assume.

4 a. Discuss the usefulness of a variable (symbol) for representing all the values that a quantity can assume.

b. Define three different variables to represent the values of each of the quantities that you defined in Task 3(a) above. (It is important when defining variables that you are specific in describing the quantity! You must include: i) the object being measured; ii) the attribute being measured; iii) where the quantity is being measured from; iv) the units of measurement.)

c. When defining a variable why is it important to describe where the quantity is being measured from?

d. When defining a variable why is it important to include the units that are used in measuring the quantity?

A **variable** is a letter that is designated to represent all possible values that a varying quantity can assume.

As an example, if we want to reference the distance (in miles) (or how far) a car has traveled from Los Angeles as it drives to Phoenix, we could define the variable *n* to represent these values (or number of miles) by writing, "let *n* represent the varying number of miles the car has driven since leaving Los Angeles and driving toward Phoenix."

---

5. Monica runs a 5-mile race around a ¼ mile track. She begins with a slow jog and ends with a sprint.
   a. Identify at least two constant quantities.

   b. Identify at least three varying quantities and state the units of measurement.

   c. Define variables to represent the values of the varying quantities you identified in part (b).

   d. Does the number of minutes it takes Monica to run 5 miles change in this situation? How many times is Monica running the 5-mile race in this situation?

   e. As the number of miles Monica has run since starting the race increases how does her distance from the finish of the 5-mile race change?

   f. Define the variable *x* to represent Monica's varying distance (in feet) from the starting line. Write an expression using *x* that represents Monica's varying distance (in feet) from the finish line. (There are 5280 feet in 1 mile.)

---

*6. Translate each of these phrases to an algebraic expression.

  a. A gas tank has 2 gallons of gas remaining when you pull into a gas station. Write an expression to represent the varying number of gallons of gasoline in the tank in terms of x, the varying number of gallons of gasoline that you add to the tank. (Illustrate this situation with a diagram first.)

  b. The city's water tank holds 20,500 gallons of water. Write an expression to represent the varying number of gallons of water remaining in the tank in terms of d, the varying number of gallons of water that drain from the full tank. (Illustrate this situation with a diagram first.)

  c. Jo is walking away from her front door in a straight line to her car that is 40 feet away. Write an expression to represent the number of feet Jo is from her car in terms of x, the varying number of feet Jo is from her front door. (Illustrate this situation with a diagram first.)

  d. Bob starts running. Bill starts running 5 seconds later. Write an expression to represent the varying number of seconds Bob has been running in terms of t, the varying number of seconds **Bill** has been running. (Illustrate this situation with a diagram first.)

$$t = Bill$$
$$t + 5 = Bob$$

  e. Bob starts running. Bill starts running 5 seconds later. Write an expression to represent the varying number of seconds Bill has been running in terms of t, the varying number of seconds that **Bob** has been running. (Illustrate this situation with a diagram first. Notice that the variable t has been redefined from part (d). Remember that it does not matter which letter we choose to represent the varying values of a quantity in some specified unit, but it is important to pay attention to precisely how variables are defined in each problem statement and/or context.)

$$t - 5$$

  f. Lisa is traveling 5 feet every second, and Sarah is traveling twice as many feet as Lisa every second. Write an expression that represents the varying total number of feet walked by Lisa and Sarah in terms of t, the varying number of seconds elapsed since the girls started walking. (Illustrate this situation with a diagram first.)

*7. View the wire on a spool animation at: https://www.rationalreasoning.net/flash/wire.html
   The spool originally has 48 feet of wire.
   a. What quantities vary in this situation?

   b. What constant values (non-varying quantities) can you identify in this situation?

   c. Define a formula to determine the remaining length of wire on the spool in terms of (or given) the amount of wire that has been cut from the spool. Define variables to represent the two varying quantities to be related first!

*8. This graph relates the depth of water in a reservoir to the number of months since January 1, 1990.

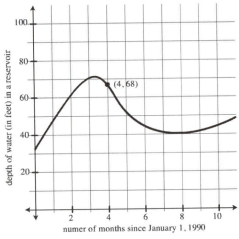

   a. Define variables to represent the values of the varying quantities in this situation.

   b. Interpret the meaning of the point (4, 68).

   c. Label 3 other points on the graph and describe what they represent.

   d. The number of months since January 1, 1990 increases from 1 to 3 months.
       i.   What is the change in the number of months since January 1, 1990?

       ii. Represent this change on the graph.

   e. Suppose the number of months since January 1, 1990 increases from 1 to 3 months.
       i.   How does the depth of the water in the reservoir vary?

       ii. What is the change in the depth of the water?

       iii. Represent the change from 1 to 3 months and the corresponding change of the water depth of the reservoir on the graph.

   g. Estimate the depth of the water in the reservoir 7 months after January 1, 1990.

   h. How does the depth of the water in the reservoir vary as the number of months since January 1, 1990 increases from 4 to 7 months? Illustrate this variation on the graph.

*9. The number of teachers in a school varies with the number of students in the school so that there are fifteen times as many students as teachers.
   a.   Identify the varying quantities in this situation.

   b.   Complete the table.

Number of students n_s in the school	Number of teachers n_t in the school
450	
600	
765	

   c.   Write a formula that defines the number of teachers in the school n_t in terms of the number of students in the school n_s.

10. For every one car in a parking lot there are 4 tires. Write a formula that defines the number of tires in the parking lot in terms of the number of cars in the parking lot. (Be sure to define variables to represent the values of the varying quantities to be related).

11. Why is it important to pick out or identify the quantities in a word problem or an applied context before trying to write a formula?

12. Rewrite each expression in an equivalent form by expanding or multiplying.

   a. $\left(\dfrac{2}{3}\right)^2 =$ b. $\left(5t\right)^2 =$ c. $12 - (3x - 4) =$

   d. $\left(\dfrac{s}{5}\right)^2 =$ e. $-5(t - 3) =$ f. $-8(-t + 7.7) =$

   g. $\left(5t\right)^3 =$ h. $-3t\,(17.5 - 5.5)^2 =$ i. $-4(-25.5s - 3) =$

1. Sam boards a Ferris wheel from the bottom and rides around several times before getting off. The following graph represents Sam's height above the ground (in feet) with respect to the amount of time (in seconds) since the Ferris wheel began moving for one complete rotation of the wheel.

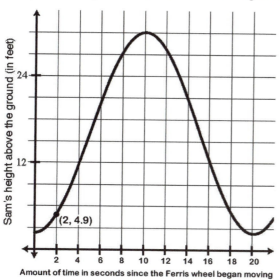

a. The point (2, 4.9) is plotted on the graph. Explain what this point conveys in the context of the Ferris wheel situation.

b. As the *number of seconds since the Ferris wheel began moving* **increased** from 2 to 8 *seconds*, how much did *Sam's height in feet above the ground* **change**? Represent this change on the graph and explain how you determined the change.

c. As the *number of seconds since the Ferris wheel began moving* **increased** from 10 to 16 seconds, what was the **change** in *Sam's height in feet above the ground*? Represent this change on the graph and explain how you determined the change.

d. Estimate the time(s) during the first 20 seconds since the Ferris wheel began moving when Sam's height above the ground was 24 feet. Represent your solution(s) on the graph and explain how you determined your solution(s).

e. Suppose we want to use *h* to represent the values of the quantity *Sam's height in feet above the ground*. What is wrong with each of the following attempts to define the variable *h*?

   i. Let *h* represent height.   ii. Let *h* represent the number of feet.

   iii. Let *h* represent the height in feet above the ground.

   iv. Let *h* represent Sam's height in feet.   v. Let *h* be Sam's height above the ground.

f. Suppose we want to use *t* to represent the values of the quantity *the number of minutes since the Ferris wheel begin moving*. Write a complete sentence that correctly defines the variable *t*.

© 2015 Carlson, Oehrtman, and Moore

2. During a baseball game, a batter hit a baseball that was later caught by an outfielder. Identify at least four quantities that are associated with this situation. You may identify both fixed and varying quantities. (Be specific in describing the attribute being measured, including where it is being measured from.)

Identify at least four quantities in this situation.

Object:	Object:
Attribute:	Attribute:
Fixed or Varying?:	Fixed or Varying?:
Choice of unit:	Choice of unit:
Object:	Object:
Attribute:	Attribute:
Fixed or Varying?:	Fixed or Varying?:
Choice of unit:	Choice of unit:

3. Translate each of these phrases to an algebraic expression.
   a. A 13-inch icicle starts melting.
      i. Let x represent the values of the quantity "the varying number of inches of the icicle that melt away." Write an expression to represent the varying number of inches of icicle remaining in terms of x. (Illustrate this situation with a diagram first.)

      ii. Let n represent the values of the quantity "the varying number of inches of the icicle remaining." Write a formula to determine n in terms of x.

b. Savanna and Porter spot each other on the beach when they are 54 feet apart. Savanna does not move, but Porter starts walking toward Savanna at a constant rate of 6 feet per second.
   i. Let *t* represent the values of the quantity "the varying number of seconds elapsed since Porter started walking toward Savanna." Write an expression to represent the varying number of feet between Savanna and Porter in terms of *t*.

   ii. Let *d* represent the values of the quantity "the varying number of feet between Savanna and Porter." Write a formula to determine *d* in terms of *t*.

c. Halle and Laura are twins. As the two girls get older (their ages vary from birth to 18 years old) their heights also vary. Halle's varying height (in inches) is represented by *x* and Laura's varying height (in inches) is represented by *y*.
   i. Use *x* and *y* to write an expression to determine how many inches taller Laura is than Halle.

   ii. As *x* and *y* vary over the first 18 years of Laura's and Halle's lives, what would it mean if the value of *y − x* was negative?

4. a. How is a formula different than an expression?

   b. The formula *n* = 13 − *x* represents the length of a melting icicle in terms of x, the varying number of inches that have melted from the 13 inch icicle, what does "*n* =" convey?

5. The original length of a candle before it is lit is 18 inches.
   a. If *x* represents the number of inches burned from the candle and *y* represents the number of inches of the candle that remain, what does 18 − *x* represent? Label *x, y,* 18, and 18 − *x* on the diagram.

   b. Write a formula that represents *y* in terms of *x*.

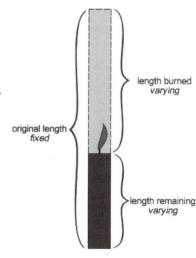

length burned
varying

original length
fixed

length remaining
varying

c. Could you have written your formula in part (b) without using a
    constant quantity? Explain.

d. What role do constant quantities play in helping you define a formula that describes how two
    varying quantities change together?

A formula defines how two varying quantities change together. Consider two varying quantities with their
varying values represented by the variables m and n.

- If we write a formula that expresses m in terms of n, we say that m is the dependent variable and n
  is the independent variable. The form of the formula is "m = [some expression with n]"

- Likewise, if we write a formula that expresses n in terms of m, we say that n is the dependent
  variable and m is the independent variable. The form of the formula is "n = [some expression
  with m]"

---

6. a. x is 2 more than y. Write a formula that defines x "in terms of" y.

    b. In the formula you created in part a, which variable is the dependent variable and which
        variable is the independent variable?

    c. Mary has been walking for t seconds and Jane has been walking for 7 seconds more than Mary.
        Define a formula to determine the varying number of seconds Jane has been walking, n, "in terms
        of" the varying number of seconds Mary has been walking, t.

    d. In the formula you created in part c, which variable is the dependent variable and which
        variable is the independent variable?

Once we have defined a variable x to represent the varying values of a quantity we can write Δx
(and say "delta x") to reference a change in x.

Remember Sam's Ferris wheel ride? If h represents Sam's height above the ground (in feet) we can write
$\Delta h = 3.2$ feet to represent a 3.2 feet *increase* (or a *positive change*) in Sam's height above the ground.
Similarly, we can write $\Delta h = -3.2$ feet to represent a *decrease* (or a *negative change*) of 3.2 feet in Sam's
height above the ground.

---

*7 a. This afternoon Jesse went for a walk to the grocery store. Let t represent the number of minutes since Jesse left his house. What is the difference in meaning of $t = 5$ and $\Delta t = 5$?

   b. Heather's electric bill in May was $56.00. Her electric bill in June was $72.50. What was the change in Heather's electric bill from May to June?

8. Fill in the tables with the specified changes in the value of the variable.

y	Δy
−7.2	
16.89	
−12.32	

m	Δm
2.1	
−3	
18.3	

s	Δs
−3	
−1.4	
5.9	

*9. At the meat counter chicken costs $3.00 a pound. When you go to the counter the butcher puts the meat on a scale before he wraps it for you to take home. As the butcher put more chicken on the scale the total weight of the meat on the scale changes. Let x represent the total weight (in pounds) of the chicken on the scale.
   a. If x increases from $x = 1$ to $x = 7$,
      i. What is the change in the value of x?

      ii. What does this change represent in this context?

      iii. What is the corresponding change in the cost of the chicken (in dollars)?

   b. If the value of x increases from $x = 13$ to $x = 19$,
      i. What is the change in the value of x?

      ii. What does this change represent in this context?

      iii. What is the corresponding change in the cost of the chicken (in dollars)?

c. How much will the cost (in dollars) of the chicken change for *any* 6-pound change in the number of pounds of chicken on the scale?

d. For the following changes in the number of pounds of chicken on the scale determine the corresponding change in the total cost (in dollars) of the chicken.
   i. $\Delta x = 2$ ii. $\Delta x = 11.7$

   iii. $\Delta x = 0.6$ iv. $\Delta x = k$ for some constant k

e. Consider the expression $x - 7$. If we substitute $x = 12$ into this expression, what are we calculating in this context?

*10. Half an hour after leaving her house Sarah was 25 miles from home. At this point Sarah merged onto the highway and was traveling at a constant speed for the next several hours. Four hours after leaving her house Sarah was 280 miles from home.
   a. In the problem we use the term "constant speed". What do you think it means for something to travel at a constant speed? Be descriptive. (That is, say more than "the speed does not change").

   b. How far did Sarah travel in 1 hour while on the highway? How do you know?

   c. How far did Sarah travel in 0.25 hours while on the highway? How do you know?

   d. How far did Sarah travel in 1.75 hours while on the highway? How do you know?

   e. How long did it take Sarah to travel 35 miles on the highway? How do you know?

   f. How long did it take Sarah to travel 1 mile on the highway?

   g. When describing constant speed we often describe it by unitizing one of the quantities involved (such as describing the speed in miles per one hour or possibly hours per one mile). What is the advantage to doing this?

11. Each of the following graphs shows two ordered pairs (x, y).

    a. Determine the value of Δx and Δy *from* the point on the left *to* the point on the right.

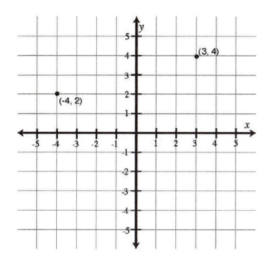 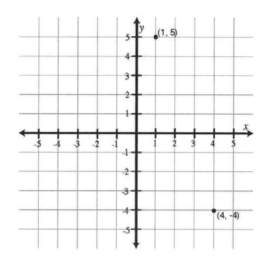

    b. Show how we can illustrate the values of Δx and Δy from part (a) on the graphs.

    c. Are your answers to part (a) different if we instead want to determine Δx and Δy *from* the point on the right *to* the point on the left? Explain.

12. Answer the following questions about the idea of *constant rate of change*.
    *a. Explain what it means for two quantities to be related by a *constant rate of change*.

    *b. David is purchasing gravel for his fish tank. The cost of the gravel increases at a constant rate of $1.25 per pound with respect to its weight. Complete the sentence: For any change in the weight of gravel purchased, the cost of the gravel (in dollars) changes....

    c. A car travels 6.5 feet along the road for each revolution of its tires, so the car's distance traveled (in feet) changes at a constant rate of 6.5 feet per tire revolution. Complete the sentence: For any change in the number of revolutions of the tires, the distance the car travels (in feet) changes...

    d. A bucket full of water has a leak. The bucket is losing water at a constant rate of 14.2 mL per minute. Complete the sentence: For any change in the number of minutes since the water started leaking out of the bucket, the amount of water in the bucket (in mL) changes...

> A **constant rate of change** exists when equal changes in the input quantity correspond to consistent changes in the output quantity.
>
> As an example if Bob is walking at a constant rate of 6.5 feet per second, any change in 1 second will always result in a change of 6.5 feet, and any change of 3.25 feet will always result in a change of ½ second.

13. Bob leaves his house and walks at a constant rate of 7 feet per second during his morning walk.
    Let *d* represent the distance (in feet) Bob has traveled since he started his morning walk
    Let *t* represent the time (in seconds) since Bob started his morning walk

    a. Determine how far Bob has walked after
        i. 0.5 seconds   ii) 0.8 seconds

    b. Verify using the points (0.5, 3.5) and (0.8, 5.6) that $\dfrac{\Delta d}{\Delta t} = \dfrac{7}{1}$ as *t* increases from *t* = 0.5 to *t* = 0.8

    c. Verify that $\Delta d = m \, \Delta t$ using all pairs of the points (0, 0) (0.5, 3.5) and (0.8, 5.6).

© 2015 Carlson, Oehrtman, and Moore

For Exercise #1 use the "Jane Walking" applet. Jane is walking from her home to work. Jane passes a mailbox when she is 25 feet from her house. Between the mailbox and a tree she walks at a constant speed, covering 40 feet in 8 seconds.

*1. Watch Jane's movement as she walks.
  a. Describe the quantities in this situation that are varying. What quantities in the situation do not vary (remain constant in the situation)?

  b. As Jane's distance from home increases, how does her distance from work change?

  c. After Jane passes the mailbox, how long does it take her to travel 20 feet? 10 feet? 5 feet? Explain your thinking.

  d. After Jane passes the mailbox, how far does she travel in 2 seconds? 6 seconds? Explain your thinking.

  e. Use the applet to help you answer the following questions.
     i.   How far will Jane travel during any 1-second time interval after passing the mailbox?

     ii. How far will Jane travel in 2.8 seconds? 3.1 seconds? k seconds?

     iii. Use your answers from part (ii) to help you write a formula to define the varying value of Jane's *change in distance* from the mailbox, Δd, in terms of the varying value of the *change in time*, Δt, since she passed the mailbox.

  f. Review your responses to parts (c) through (e). Describe what it means for an object to move at a constant speed. (*Note: Say something more than "The speed doesn't change" – be descriptive and reference specific quantities.*)

In Exercise #1 we saw examples where as the values of two quantities x and y changed together, the change in one variable was always some constant m times as large as the change in the other variable. This fixed relationship that describes how x and y change together describes what it means for one quantity to change at a **constant rate of change** with respect to another quantity.

There exists a **constant rate of change** between two quantities (call them A and B) if the below statement is true for all corresponding values of Quantity A and Quantity B:

$$\text{(change in the value of Quantity B)} = \text{(constant number)} \cdot \text{(change in the value of Quantity A)}$$

We call the value of the constant number the *constant rate of change* of Quantity A with respect to Quantity B, and its value can be calculated as follows:

$$\text{constant rate of change} = \frac{\text{change in the value of Quantity B}}{\text{change in the value of Quantity A}}$$

Suppose *x* is the value of Quantity A and *y* is the value of Quantity B, then **y changes at a constant rate with respect to x if, as x and y change together,**
   i) Δy is always *m* times as large as Δx (We can also write $\Delta y = m \cdot \Delta x$)
   ii) the ratio of the change in y to the change in x is always the same constant, *m*
   (We can also write $m = \frac{\Delta y}{\Delta x}$)

*2. Each of the following represents a situation with a constant rate of change. Use the definition above to explain what these statements mean:
   a. $\Delta d = 6 \cdot \Delta t$ (Hint: As the values of *x* and *y* change together, the change in *y* is always….)

   b. $\Delta y = -2 \cdot \Delta x$

   c. $\Delta y = \frac{1}{2} \cdot \Delta x$

   d. Illustrate the meaning of $\Delta y = -2 \cdot \Delta x$ on a graph that also has the point (1, 7).

   e. As x varies from 1 to 5.5, how does *y* vary? Illustrate these variations on your graph.

   f. Does *x* take on all values between 1 and 5.5? If so, how is this illustrated on your graph?

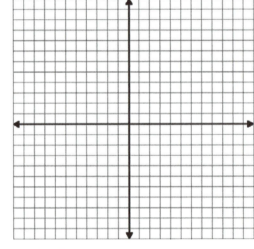

*3. Suppose that the cost of purchasing a certain type of landscaping rock is $107 per ton.
   a.   What two quantities vary in this situation? What quantity is constant?

   b.   Determine the constant rate of change *m* (if one exists) of the *total cost of the landscaping rock* with respect to the *number of tons of landscaping rock purchased.*

   c.   Write a formula to determine the **change in the total cost** of landscaping rock purchased with respect to the **change in the number of tons** of landscaping rock purchased.

d. Write a formula to determine the *cost of the landscaping rock purchased* in terms of the *number of tons of landscaping rock purchased.* What do you notice about this formula and the formula you defined in part (c)? Explain.

*4. Suppose that the values of *y* change at a constant rate with respect to the values of *x*.
   a. Given the information in this table,

x	−2	0	5
y	8	2	−13

   i. determine the value of *m*, the constant rate of change of *y* with respect to *x*.

   ii. write a formula to express how Δy and Δx are related (that expresses Δy in terms of Δx).

   iii. write a formula to express how *y* and *x* are related (that expresses *y* in terms of *x*).

   b. Given the information in this graph,

   i. determine the value of *m*, the constant rate of change of *y* with respect to *x*.

   ii. write a formula to express how Δy and Δx are related (that expresses Δy in terms of Δx).

   iii. write a formula to express how *y* and *x* are related (that expresses *y* in terms of *x*).

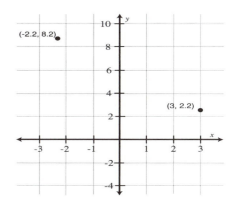

*5. Jane is now walking from her house to the store and passes her mailbox which is 100 feet from her house. Let *d* represent Jane's distance *from her house*, measured in feet, and let *t* represent the time in minutes that elapsed since Jane *passed her mailbox.* The table below shows the distance Jane is from her house at various times since she passed her mailbox.

Time in minutes	0	5	12	26
Distance in feet	100	220	388	724

   a. There are three pairs of points in this table: (0 , 100) and (5 , 220); (5 , 220) and (12 , 388); (12 , 388) and (26 , 724). Compute Δt and Δd for each of these pairs, and then compute the ratio $\dfrac{\Delta d}{\Delta t}$ for each pair. What do you notice?

   b. Compute Δt and Δd for the pair of points (5, 220) and (26, 724), and then compute the ratio $\dfrac{\Delta d}{\Delta t}$. What do you notice?

   c. Based on the information in the table is it possible that Jane was walking at a constant rate of change? Explain your answer.

d. Assume that Jane is walking at a constant rate of change. Based on the information in this table, construct a formula that expresses Δd in terms of Δt.

e. Construct a formula that expresses d in terms of t (Hint: Remember that d represents Jane's distance from her house, NOT her mailbox.)

f. Construct a graph that represents Jane's distance from her house d, measured in feet in terms of the number of minutes t since she passed her mailbox. Illustrate Δt and Δd on your graph for two pairs of points.

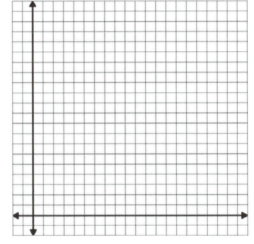

> If there is a constant rate of change of one quantity with respect to another quantity, then we say that the relationship is a **linear function**.

*6. Given that the x and y axes have the same scale for both graphs below, answer the following questions.

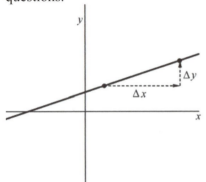

Graph of linear function A

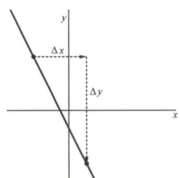

Graph of linear function B

a. What is the approximate constant rate of change of y with respect to x for the linear functions whose graphs are given?
Linear function A: Linear function B:

b. Write a formula that expresses Δy in terms of Δx.
Linear function A: Linear function B:

c. Write a formula that expresses y in terms of x given the y-intercept of the graph of function A is $(0, 3)$ and the y-intercept of the graph of function B is $(0, -3)$.

Formula for linear function A: Formula for linear function B:

*7. We have established that as x and y change together, there exists a constant rate of change of y with respect to x if $\Delta y = m \cdot \Delta x$ (that is, if the change in y is always m times as large as the change in x, **as x and y vary together**).

 a. Suppose that y changes at a constant rate of ¾ with respect to x. As x and y vary together, for any change in x of 3 ($\Delta x = 3$) in this relationship we can say that the change in y (Δy) is equal to what?

 b. Restate part (a) in your own words.

 c. Given that the point (1, 2) is on the graph of this linear function, illustrate Δy for 2 different instances when $\Delta x = 3$.

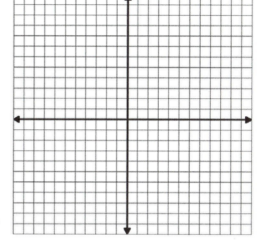

*8. Given that a candle is 8.5 inches tall when it has been burning for 6 hours at a rate of ¼ in. per hour,
 a. how tall was the candle when it had been burning for 2.4 hours.

 b. how tall was the candle before it was lit (when it had been burning for 0 hours)?

9a. Graph the line $y = 4x$ and plot 3 points on your graph, then illustrate Δy and Δx between (0, 0) and each of these points.

 b. Consider the graph that you constructed in (a). For each point (x, y) on your graph, compare the values of x and y to the values of Δx and Δy between (0, 0) and (x, y). What do you notice?

 c. Given that $\frac{\Delta y}{\Delta x} = 4$ for some linear function, under what conditions does $\frac{y}{x}$ also equal 4?

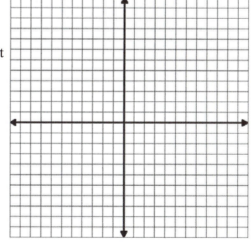

*10. A slow running hose is used to fill an empty wading pool. The graph represents the volume of water (in gallons) in the pool, *v,* in terms of the amount of time elapsed, *t,* (in minutes) since the pool began filling.

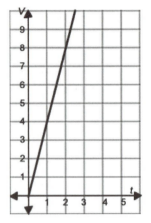

a. Write a formula that determines the volume of water, *v,* (in gallons) in terms of the time elapsed, *t,* (in minutes) since the pool began filling?

b. How many times as large is Δv than Δt ? Express this relationship with a formula.

c. How are the formulas you wrote in parts (a) and (b) related?

d. Write a formula that determines the elapsed time, *t,* (in minutes) since the pool began filling in terms of the volume of water, *v,* (in gallons) of water in the pool.

e. How many times as large is Δt than Δv? Express this relationship with a formula and explain how it relates to the formula you wrote in part (d).

f. What do you notice about how your formula in part (a) and part (d) are related? Do they represent the same relationship? Explain.

\

*1. The perimeter p of a square (measured in inches) is expanding in size at a constant rate of change of 4 inches per inch with respect to the length of the square's side s (measured in inches).

   a. If the length of the side of the square changes (increases) from
       i.   0 to 2.5 inches, how does the square's perimeter p change? (Use your formula from part (a) to represent this change.)

       ii. 1 to 3 inches, how does the square's perimeter p change?

   b. Express the **change** in the square's perimeter p in terms of the **change** in the square's side length s.

   c. Write a formula and construct a graph that represent the perimeter of the square p (in inches) in terms of the square's side length s (in inches).

       Formula: _____

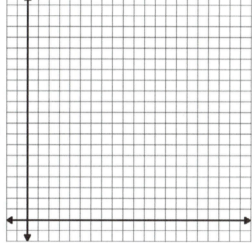

   d. Represent the changes you computed in part (a) on the graph and use the graph to verify that the formula you wrote in part (b) holds for the relationship between the changes in p and s.

If y varies with x in a linear relationship so that $y = mx$ ("the value of y is m times as large as the value of x"), then we can think of any value of x or y as a change from 0. It is then also true that $\Delta y = m\Delta x$ ("the change in y is m times as large as the change in x, as the values of x and y change together").

---

2. A framer has asked you to write a formula to determine the length of wood board needed to build a frame for square pictures with varying side lengths. Since some wood will be wasted when making the corner cuts the framer wants the board to be 8 inches longer than the picture's perimeter.

   a. What quantities change together in this situation?

   b. If your goal is to write a formula to relate the square's side length to the length of board to be cut for the framer,
       i.   what two varying quantities are you being asked to relate?

       ii. Define variables to represent the values of these quantities.

       iii. Now write a formula that defines the length of board to be cut (in inches) *in terms of* the side length of the square picture (inches):

---

c. Write a formula that defines the *change* in the length of the board to be cut (in inches) in terms of the *change* in the picture's side length (in inches).

    i. Formula: _____

    ii. As the side length of various pictures changes from 4 inches to 12 inches, how does the length of the board to be cut (in inches) change?

When defining the values of one quantity, *y*, *in terms of* the values of another quantity, *x*
- *x* represents the values of the independent quantity, typically represented by the horizontal axis
- *y* represents the values of the dependent quantity, typically represented by the vertical axis

(Representing the values of *y* with the horizontal axis and the values of *x* with the vertical axis would not change how *x* and *y* are related, provided the axes are also relabeled. However, using consistent practices and conventions when graphing can improve communication.)

---

*3. Illustrated here are two cylinders: one wide and one narrow. Both cylinders have equally spaced marks for measurement. Water is poured into the wide cylinder up to the 4th mark (see A). This water rises to the 6th mark when poured into the narrow cylinder (see B).

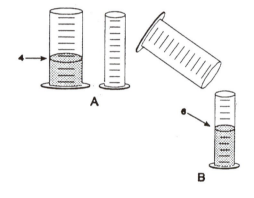

  a. Both cylinders are emptied, and water is poured into the narrow cylinder up to the 11th mark. How high would this water rise if it were poured into the empty wide cylinder?

  b. Imagine pouring water into the narrow cylinder and let *x* represent the varying mark number the water reaches in the narrow cylinder as you pour. Let *y* represent the mark number the water would reach if we poured the same volume of water into the wide cylinder.

    i. What is the meaning of $\frac{y}{x} = \frac{2}{3}$ in this context?

    ii. What is the meaning of $y = \frac{2}{3}x$ in this context?

    iii. What is the meaning of $\Delta y = \frac{2}{3}\Delta x$ in this context?

  c. i. When *x* = 1, what is the value of *y?*

    ii. When *x* = 3, what is the value of *y?*

    iii. When *x* = 6, what is the value of *y?*

    iv. When *x* = 4.5, what is the value of *y?*

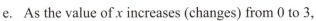

d. Plot the four points (x, y) you determined in part (c) on the given axes. Be sure to label your axes with the quantities whose values you are comparing.

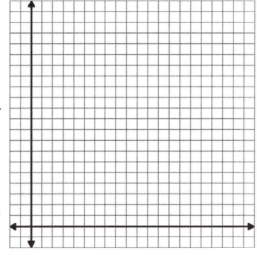

e. As the value of x increases (changes) from 0 to 3,

   i. what is the value of Δy (what is the change in the mark number the water reaches in the wide cylinder as the mark number the water reaches in the narrow cylinder changes from 0 to 3)? Illustrate these changes on your graph.

   ii. what is the value of y when $x = 3$ (also thought of as a change of positive 3 units from an initial value of 0 units)? How are these values illustrated on your graph?

f. As the value of x increases (changes) from 3 to 6, what is the value of Δy? Illustrate these changes on your graph.

g. Discuss how $\dfrac{y}{x} = \dfrac{2}{3}$, $y = \dfrac{2}{3}x$, and $\Delta y = \dfrac{2}{3}\Delta x$ are illustrated on the graph, and remind yourself of the meaning conveyed by each formula.

4. The graph of a certain linear relationship with a constant rate of change of y with respect to x of -2.5 passes through the point $(-5, 8)$.

   a. Plot the given point on the graph, then use the constant rate of change to determine the vertical intercept. Explain the thinking you used to determine your answer.

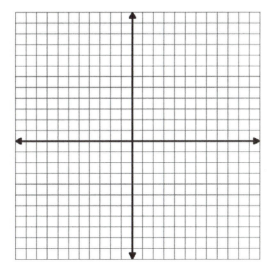

   b. Use the given point and the constant rate of change to complete the table of values for this relationship.

x	y
-6	
-0.8	
0	
2.3	

*5. The length of a burning candle decreases at a constant rate of
2.2 inches per hour. The candle has been burning for 3.5 hours
and is currently 8.3 inches long (the point (3.5, 8.3) on the
graph represents these two co-occurring values of the quantities).

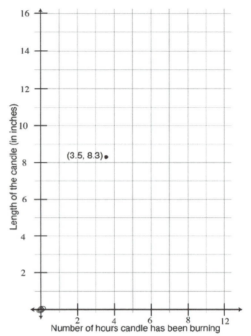

a. Represent "an increase in the time spent burning of 2 hours"
from the given point on the graph.

b. How much will the length of the candle (in inches) change
when the time spent burning increases by 2 hours?
Represent this on the graph.

c. Represent "a decrease in the time spent burning of 1.8 hours"
from the given point on the graph.

d. How much will the length of the candle (in inches) change
when the time spent burning is decreased by 1.8 hours?
Represent this on the graph.

e. What is the change in the length of the candle if the candle
has been burning for *x* hours?

f. What was the original length of the candle (in inches) before it started burning? Explain how you
determined this value and represent your reasoning on the graph.

g. On the same axes draw the graph that represents the length of the candle in inches in terms of the
number of hours spent burning.

6. A 19-inch candle is lit. The candle burns away at a constant rate of 2.5 inches per hour.
a. Draw a diagram that represents a "snapshot" in the process
of the burning candle. To the side of your diagram, identify
every quantity that is relevant to the process. Define variables
to represent the values of the varying quantities that you identify.

b. Represent the variables you defined in part (a) on your diagram,
being careful to indicate accurately where measurement of the
quantities starts. You may need to use arrows or line segments
to show how the variables relate to the candle and to each other.

c. Write a formula to represent the change in the length of the
candle (in inches) in terms of the change in the number of
hours since the candle was lit.

d. How much time will it take to burn the entire candle?
Explain your reasoning.

e. Write a formula to represent the length of the candle (in inches) *in terms of* the number of hours
since the candle was lit.

© 2015 Carlson, Oehrtman, and Moore

*7. Alan and Alissa are attempting to find each other in a crowded mall. They finally spot each other near the food court, at which point they are 140 feet apart. They begin walking towards each other. Alan travels at a constant rate of 5 feet per second and Alissa travels at a constant rate of 4 feet per second.

    a.   What quantities are varying in this context? What quantities are constant? Be sure to include units for each quantity. Draw a diagram of the situation if useful.

    b.   Complete the following table of values.

number of seconds since Alan and Alissa spotted each other	distance (in feet) Alan has traveled	distance (in feet) Alissa has traveled	distance (in feet) between Alan and Alissa
0	0	0	140
1			
2			
4			
8			
10			

    c.   For each second that passes since Alan and Alissa spotted each other, how many feet closer are Alan and Alissa? What about the situation helps to explain this?

    d.   Define formulas to relate each of the following quantities. Be sure to clearly define any variables you use.

        i.   Alan's distance traveled (in feet) in terms of the number of elapsed seconds since spotting Alissa.

        ii.   Alissa's distance traveled (in feet) in terms of the number of elapsed seconds since spotting Alan.

        iii.   The distance between Alan and Alissa (in feet) in terms of the number of elapsed seconds since they spotted each other.

    e.   Use your formula(s) created in part (d) to determine how many seconds it will take for Alan and Alissa to reach one another.

8. A tortoise and hare are competing in a 1600-meter race. The arrogant hare decides to let the tortoise have a 950-meter head start. When the start gun is fired, the hare begins running at a constant speed of 9.5 meters per second and the tortoise begins crawling at constant speed of 4 meters per second.

    a. Illustrate the situation with a diagram, and then define a formula to determine the varying distance between the tortoise and the hare in terms of the number of seconds since the start gun was fired. Be sure to define your variables before trying to write the formula.

    b. Construct a graph to represent the varying distance between the tortoise and hare in terms of the number of seconds since the start gun was fired.

    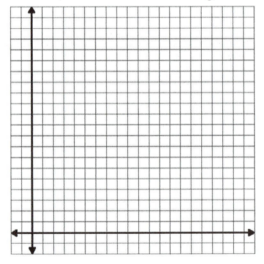

    c. Who finishes the race first, the tortoise or the hare? How long does it take each animal to complete the race? Explain.

© 2015 Carlson, Oehrtman, and Moore

1. A car is driving away from a crosswalk. The distance d (in feet) of the car from the crosswalk t
    seconds since the car started moving is given by the formula $d = t^2 + 3.5$.
    a. As the number of seconds since the car started moving increases from 1 second to 3 seconds,
        what is the change in the car's distance from the crosswalk?

    b. Illustrate on the axes: i) t increasing from 1 to 3 seconds,
        ii) the corresponding change in the car's distance from the
        crosswalk as the value of t increases from 1 to 3 seconds.

    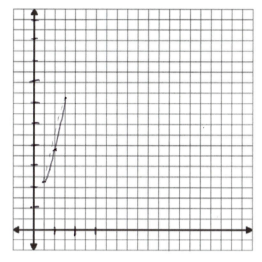

    c. True or False: The car travels at a constant speed as the
        value of t increases from 1 to 3 seconds.
        Discuss and explain. (Hint: It may help to think about how
        far the car travels in the 1st second compared to how far it
        travels in the 2nd second).

    d. Suppose we want to estimate the speed of the car over this interval from $t = 1$ to $t = 3$. We could $\dfrac{12.5 - 4.5}{3 - 1}$
        determine what constant speed would have resulted in the car traveling the same distance (as the
        distance that the car actually traveled) over this time interval from $t = 1$ to $t = 3$ seconds. $\boxed{4} \frac{8}{2}$
        i)   Connect the two points with a straight line on your graph in part (b),
        ii)   Determine the slope of this line and describe what this slope represents in the context of
                 this question.

        iii)   True or False: The straight line segment represents the graph of the actual distance (in
                 feet) of the car from the crosswalk as t increases from 1 to 3 seconds. Explain.

    e. Using the formula $d = t^2 + 3.5$, determine the value of d when $t = 0, 1, 1.25, 1.8, 2, 2.5, 3, 4$

t	0	1	1.25	1.8	2	2.5	3	4
d	3.5	4.5	5.06	6.74	7.5	9.75	12.5	19.5

    f. Sketch the graph of $d = t^2 + 3.5$ on the same axes in part (b). What does this graph represent in
        the context of this problem?

    g. Illustrate on the same axes in part (b) above: i) t increasing from 3 to 4 seconds, ii) the
        corresponding change in the car's distance (in feet) from the crosswalk as the value of t increases
        from 3 to 4 seconds, and then (iii) connect the two points with a straight line and describe what
        the slope of this line represents.

2. The distance d (in feet) of a car north of an intersection t seconds after it started moving is given by the formula $d = 2t^2 - 3$. (Note, we use number subscripts on d and t to show corresponding pairs of these variables' values—meaning t_0 is a value of t that corresponds to a specific value, d_0, of d.)

   a. Determine the value of d_1 when $t_1 = 2$

   b. Determine the value of d_2 when $t_2 = 3.5$

   c. Explain what $\dfrac{d_2 - d_1}{t_2 - t_1}$ represents in the context of this situation (Hint: What constant rate of change is this?)

   d. Construct a graph of $d = 2t^2 - 3$ and then illustrate the constant speed a car would have needed travel in order to cover the same distance in the same amount of time as t increases from $t_1 = 2$ to $t_2 = 3.5$.

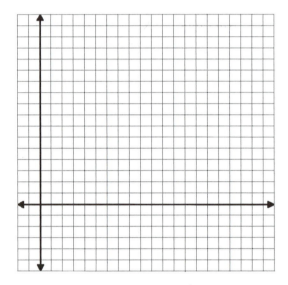

Constant Rate of Change Over an Interval of the Independent Quantity: A formula of the form $y = $ <some expression that contains x> describes how the values of y (the dependent quantity) change in terms of varying value of x (the independent quantity). The graph of the formula represents how the values of the independent and dependent quantities change together. The constant rate of change of y with respect to x over any interval of the independent quantity from x_1 to x_2 is $\dfrac{y_2 - y_1}{x_2 - x_1}$. This constant rate of change is represented graphically as the slope m of the straight line connecting the points (x_1, y_1) and (x_2, y_2).

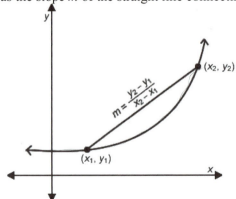

3. When running a road race you heard the timer call out 8 minutes as you passed the first mile-marker in the race.
   a. What quantities will you measure to determine your speed as you travel? Define variables to represent the quantities' values and state the units you will use to measure the value of each of these quantities.

   b. As you passed mile-marker 6 you heard the timer call out 52 minutes. *If you had* held a constant speed (rate of change) as you ran from mile marker 1 to mile marker 6, determine the constant speed (in miles per minute) you *would have* ran as you traveled from mile marker 1 to mile marker 5.

   c. After mile-marker 6 you slowed down and ran at a constant speed of 10 minutes per mile between mile-marker 6 and mile-marker 10, how many minutes did it take you to travel from mile marker 6 to 9?

4. Marcos traveled in his car from Phoenix to Flagstaff, a distance of 155 miles.
   a. Determine the amount of time required for Marcos to travel from Phoenix to Flagstaff if he drove at a constant speed of 68 miles per hour.

   $$155m \times \frac{1h}{68m} = 2.28 \text{ hours}$$

   b. Construct a graph to represent Marcos's distance from Phoenix in terms of the time (in hours) since he left Phoenix.

   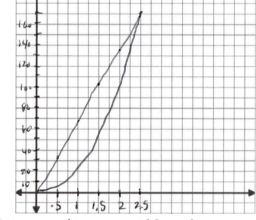

   c. Joni left Phoenix at exactly the same time as Marcos and arrived in Flagstaff at exactly the same time as Marcos, **but did not drive at a constant speed.**

      i) <u>True</u> or False: Joni covered the same distance in the same amount of time as Marcos. Explain.

      ii) Construct a possible graph (on the same axes in part (b) that you used to construct Marcos's distance-time graph) to represent Joni's distance traveled in miles in terms of the number of hours since she left Phoenix. Discuss the thinking you used to construct your graph.

      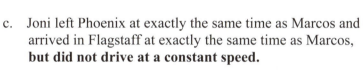
      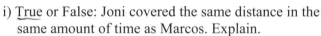

5. Suppose that it took your car 18 minutes to travel from Kansas City to mile marker 363 and 30 minutes to travel from Kansas City to mile marker 373.
   a. Illustrate this situation with a drawing.

b. We have seen that determining the constant speed of an object over an interval of time is determined by examining the amount of distance traveled during that interval of time. At what constant speed (in miles per minute) did your car travel as it moved from mile marker 363 to mile marker 373? Illustrate how you computed your answer and describe the rationale for your approach.

$$\frac{10}{30} = .333 \text{ or } \frac{1}{3} \text{ mile per min}$$

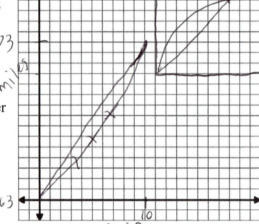

c. Construct a possible distance-time graph to display how your car's distance (in miles) from mile marker 363 to 373 varied with the number of minutes elapsed after passing mile marker 363. Label your axis and label the points on your graph that represent the car's starting and ending positions.

d. A truck passed your car at mile marker 363. At mile marker 373, your car passed the truck. On the same axes on which you just graphed the distance-time relationship of your car as it traveled from mile marker 363 to mile marker 373, construct a possible distance-time graph of the truck's distance from mile marker 363 at it traveled to mile marker 373.

e. What do the intersection points (where the two graphs intersect) on your axes convey about the relative distance traveled by the car and truck, at specific times (numbers of seconds) since the car left Kansas City? The car/Truck traveled the same distance in the same time.

f. Using the graphs that you created above, determine the distance between your car and the truck 6 minutes after the vehicles passed mile marker 363.

g. Was the truck's speed between mile markers 363 and 373 ever exactly the same as your car's speed? Explain.

h. If the truck continues traveling in the same direction, now at the constant speed as the car, how many minutes will it take the truck to drive 45 miles from mile marker 373?

*6. The distance d (in feet) of a car north of an intersection, t seconds since the car started to move, is given by the formula $d = 1.5t^2 + t + 3$.

a. As the time t since the car started to move increases from $t = 3$ seconds to $t = 7$ seconds, what constant speed would a truck need to travel to cover the same distance over this 4-second time interval as the car? $\frac{83.5 - 19.5}{7 - 3} = 78.62$ ft/s

b. The constant speed needed to travel some distance over some interval of time is often called the average speed (or *average rate of change*). What is the meaning of average rate of change in the context of this problem?

1. You are considering buying a house that is located at the intersection of D Street and 23rd Avenue. Your cell phone provider is building a cell phone tower at the intersection of B Street and 20th Avenue. Each street is located one mile apart. If you are farther than 3.5 miles away from the cell tower, you will not have good reception.

   a. What quantity do you need to find to determine if you will have good reception at the house? Illustrate these quantities on the map.

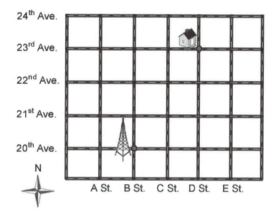

   b. Use the map to measure the quantities that are needed to determine the distance of the cell tower from the house to decide if the house will have good cell coverage. Label these quantities on the map. (Hint: You will need to use the Pythagorean Theorem to compute the distance of the tower from the house.)

   c. Define two variables (that represent the values of two quantities) that can be used to determine the distance of *any house* from the tower.

   d. Use the variables defined in part (c) to determine a formula for finding the distance of *any house* from the tower.

   e. If a house is located at (x_1, y_1), modify the formula in part (d) to determine the distance *d* of the house from the cell tower.

2. Consider a cell phone tower whose location is represented by the point $(-2, 4)$ on the coordinate plane and whose radius of cell coverage is 3 miles. Note that every point 3 miles away from the cell tower will be on the boundary of the cell coverage. This will form a circle around the point $(-2, 4)$ with a radius of 3 units, as shown in the given figure.

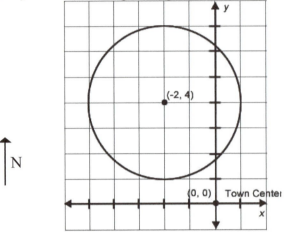

    a. A point $(x, 2)$, southeast of the tower is on the service boundary of this cell tower. What is the value of x?

    b. Let (x, y) be any point on the boundary of the cell tower. Alter the formula you created in part (1e) to write a formula that represents the distance of any house from the cell tower located at $(-2, 4)$.

*1. A carpenter needs to cut a wooden board to a length of 7.6 inches with a **tolerance** of 0.1 inch, meaning the actual length after the cut can be within 0.1 inch of 7.6 inches and still be usable.
   a. Let x represent the actual length of the board after cutting. List some possible values for x that represent usable board lengths, and then represent all possible values on the number line.

   b. Write a mathematical statement (inequality) to represent the possible usable board lengths.

   c. We will call the difference between the actual board length and the desired board length the **error**. Write an expression to represent the error.

   d. Write a mathematical statement (inequality) that represents errors within our tolerance of 0.1 inch.

   e. Write a mathematical statement (inequality) that represents actual board lengths x that are not usable, then represent these values on a number line.

*2. Suppose we want to choose an x value within 4.5 units of 16.
   a. List some values of x that meet this constraint.

   b. Write a mathematical statement (inequality) to describe the possible changes in x away from 16.

   c. Write a mathematical statement (inequality) to describe the possible values of x.

   d. Represent the possible values of x on the number line.

3. Suppose we want the change in x away from 5 to be positive but no more than 3.
   a. List some values of x that meet this constraint.

   b. Write a mathematical statement (inequality) to describe the possible changes in x away from 5.

   c. Write a mathematical statement (inequality) to describe the possible values of x.

   d. Represent the possible values of x on the number line.

4. Suppose we want to choose a value of x that is at least 4 units away from $x = -3$.
   a. List some values of x that are at least 4 units away from $x = -3$, then represent all possible values on the number line.

   b. Write a mathematical statement (inequality) that describes values of x at least 4 units away from $x = -3$. Try to think of at least two different ways to write this.

Definition of Absolute Value

Given a real number $c > 0$,

$|x - h| = c$ represents all values of x that are exactly c units away from h on the real number line

$|x - h| < c$ represents all values of x that are less than c units away from h on the real number line

$|x - h| > c$ represents all values of x that are greater than c units away from h on the real number line

---

5. Let's revisit your answers to the previous exercises. If you didn't already, use absolute value to rewrite the solutions to Exercises #1c, #2b, and #4b.

*6. Jamie is assembling a model airplane that flies. The instructions say that the best final weight is 28.5 ounces. However, the actual weight can vary from this by up to 0.6 ounces without affecting how well the model flies.

  a. Represent the weights that are "acceptable" on the number line and describe them using an inequality.

  b. Write a mathematical statement (inequality) that represents the acceptable variation in the model's weight
   i. using absolute value.   ii. without using absolute value.

7. a. What is the meaning of the statement "$|x-5| < 2$"?

  b. Represent all values of x that make the statement true on the number line.

  c. Rewrite the statement in part (a) without using absolute value.

8. a. What is the meaning of the statement "$(w-1) \geq 5$ or $(w-1) \leq -5$"?

  b. Represent all values of w that make the statement true on the number line.

  c. Rewrite the statement in part (a) using absolute value.

*9. Use absolute value notation to represent the following.

    a. All numbers x whose distance from 6 is no more than 7.5 units.

    b. All numbers r whose distance from –2 is more than 1 unit.

*10. For each statement, do the following.

    i. Describe the meaning of the statement.

    ii. Rewrite the statement without using absolute value.

    iii. Determine the values of the variable that make each statement true.

    iv. Represent the solutions on a number line.

    a. $|x-11|<6$ b. $|y+7|\geq1$

    c. $|a-3.3|>0.8$ d. $|p+2|\leq5$

11. a. Use the definition of absolute value to describe the solutions for each of the following statements, then give the values of x that make the statement true.

    i. $|x-0|=6$

    ii. $|x-0|<3$

   b. Based on your work in part (a), what does each of the following represent for a real positive number c?

    i. $|x|=c$

    ii. $|x|\geq c$

12. For each statement, do the following.
    i.   Describe the meaning of the statement.
    ii. Write the statement without using absolute value.
    iii. Represent the solutions on the number line.

a. $|x| = 4$

b. $|x| \le 5.5$

c. $|x| > 3$

I.   QUANTITIES AND CO-VARIATION OF QUANTITIES (Text: S1)

1. Consider a cup of coffee.
   a.   Identify five attributes of the cup of coffee that can be measured. Which of these attributes are fixed and which are varying?
   b.   Identify five attributes of the cup of coffee that cannot be measured.

2. You just got home from a trip to the grocery store and identify some elements and aspects of your trip. Determine if the following statements define a quantity. If not, rewrite the statement so that the statement properly defines a quantity.
   a.   the bag of groceries
   b.   the number of apples purchased at the grocery store
   c.   the money spent at the grocery store
   d.   the gasoline used to get to the grocery store
   e.   the grocery store

3. A group of students are taking an exam. You walk into the room and identify some elements and aspects of the situation. Determine if the following statements define a quantity. If not, rewrite the statement so that the statement properly defines a quantity.
   a.   the number of people taking the exam
   b.   the people sitting around a table
   c.   the exam
   d.   the questions on the exam
   e.   the teacher

4. For the following situations identify one constant quantity and two varying quantities. Define variables to represent the values of the varying quantities.
   a.   A mountain climber hikes with two friends for 5 hours.
   b.   The computer charges for 1.25 hours each night.
   c.   The student studies for 8 hours each weekend.
   d.   Jessica bikes 30 miles around a 5 mile course.

5. You fill up a bottle of water in the sink. You do not turn off the water until the bottle is full.
   a.   Identify at least two constant quantities in this situation.
   b.   Identify at least two varying quantities in this situation and state the units of measurement.
   c.   Define variables to represent the values of the varying quantities you identified in part (b).
   d.   As the number of minutes since you began filling up the water bottle increases how does the height of the water in the water bottle change?

6. For the following situations identify the quantities whose values vary and the quantities whose values are constant. State possible units for measuring each of these quantities. Then define variables to represent the values of each varying quantity.
   a.   A 10-inch candle burns for 2 hours.
      i.   Identify at least one constant quantity and state the units of measurement.
      ii.   Identify at least two varying quantities and state the units of measurement.
      iii. Define variables to represent the values of the varying quantities you defined in part (ii).
   b.   A stone layer installs square tiles that have a 9-inch diagonal length on patio floors.
      i.   Identify at least one constant quantity and state the units of measurement.

(problem continues on next page)

      ii. Identify at least two varying quantities and state the units of measurement.

      iii. Define variables to represent the values of the varying quantities you defined in part (ii).

  c. A girl runs around a ¼-mile track.

      i.   Identify at least one constant quantity and state the units of measurement.

      ii. Identify at least two varying quantities and state the units of measurement.

      iii. Define variables to represent the values of the varying quantities you defined in part (ii).

  d. A scuba diver descends from the surface of the water to a depth of 60 feet.

      i.   Identify at least one constant quantity and state the units of measurement.

      ii. Identify at least two varying quantities and state the units of measurement.

      iii. Define variables to represent the values of the varying quantities you defined in part (ii).

7. The number of questions on an exam varies with the number of minutes to take the exam. For each question there are 5 minutes allotted. Write a formula that relates the number of minutes to take the exam to the number of questions on the exam. (Be sure to define variables to represent the values of the varying quantities.)

8. There are twelve times as many football players on a football team as there are coaches. Write a formula that relates the number of coaches to the number of players on the team. (Be sure to define variables to represent the values of these varying quantities).

9. The graph below relates the total value of world exports (internationally traded goods) in billions of dollars to the number of years since 1950.

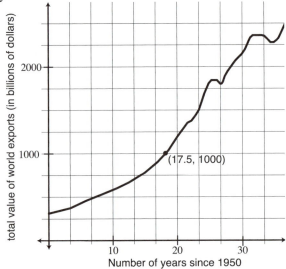

  a. Define variables to represent the values of the varying quantities in this situation.

  b. Interpret the meaning of the point (17.5, 1000).

  c. As the number of years since 1950 increases from 10 to 20 years what is the change in the number of years since 1950? Represent this change on the graph above.

  d. As the number of years since 1950 increases from 10 to 20 years what is the corresponding change in the total value of world exports (in billions of dollars)? Explain how you determined this value and represent this change on the graph above.

10. A ball is dropped off of the roof of a building. The graph below relates the height of the ball above the ground (in feet) to the number of seconds that have elapsed since the ball was dropped.

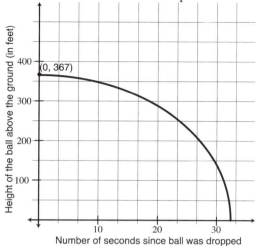

a. Define variables to represent the values of the varying quantities in this situation.
b. Interpret the meaning of the point (0, 367).
c. As the number of seconds since the ball was dropped increases from 10 to 30 seconds, how does the height of the ball above the ground change?

11. Write an equation for each of the relationships that are described below. Then solve the equation for the unknown. (Do not forget to start by defining a variable to represent the unknown value.)
a. 17.5 is equal to 2 times some number. What is the number?
b. The sum of 3 times some number and 12 is 42. What is the number?
c. ¼ of some number is 4.3. What is the number?
d. Some number is 4 times as large as 9.8. What is the number?
e. Some number is equal to 1/3 of the sum of 88.2, 93.5, and 64. What is the number?
f. The change from some number to 12.5 is – 5. What is the number?
g. Measuring some number in units of 12 is 3.5. What is the number?

12. Write an equation for each of the relationships that are described below. Then solve the equation for the unknown. (Do not forget to start by defining a variable to represent the unknown value.)
a. 45 is some multiple of 15. Determine the value of the multiple.
b. $1,200 is 1.5 times as large as some amount of money. What is the amount of money?
c. 200% of some number is 38.2. What is the number?
d. 5 is the result of 12 more than ¾ of some number. What is the number?
e. Suppose 10 is the number that is 5 times as large as the value that is 4 less than the value of x. What is the value of x?
f. The ratio of the change from 2 to 8 and the change from 5 to 7. What is the ratio?
g. 7 is 3.5 times as large as some number. What is the number.

13. Evaluate the following expressions:
a. $-2 + 5 - 12$
b. $6 - (-4) + 1$
c. $2(-3)(-1) - (-6)$
d. $\dfrac{-5 + (9 - 5(4)) + 3}{5(-2) - 1}$

14. Let $y = 3.5x - 7$.
    a. Find the value of y when x is zero.
    b. What value(s) of x give a y value of 11?
    c. What value(s) of y correspond to an x value of 3?

15. Solve each of the equations for the specified variable.
    a. Given $y = 17x - 6$, solve for x when $y = 45$.
    b. Given $y = \frac{6x+5}{3}$, solve for x when $y = 2$.
    c. Given $z = \frac{12x-4}{3}$, solve for x when $z = 10$.

    d. Given $y = \frac{3.2x(6-0.5x)}{x}$, solve for x when $y = 1$.
    e. Given $y = \frac{-6a+7-2a}{2} - 3 + a$ solve for a when $y = 2$.

16. Simplify the following
    a. $\dfrac{3x^3 + 6x}{x}$
    b. $\sqrt{52x^8 y^3}$
    c. $\dfrac{2x^2 + x - 6}{x + 2}$

17. Simplify the following
    a. $\dfrac{4(x+3) + 6x - 12}{x}$
    b. $\sqrt{70x^9 y^{153}}$
    c. $\dfrac{6x^2 - 5x - 21}{(2x+3)(3x-7)}$

II. CHANGES IN QUANTITIES AND CONSTANT RATE OF CHANGE (Text: S1, 3)

18. Matthew started working out. As a result his weight went from 175 pounds to 153 pounds. What was the change in Matthew's weight?

19. After driving from Tucson to Phoenix the number of miles on your car's odometer went from 312 to 428. What was the change in the number of miles on your car's odometer?

20. After spending an hour processing emails the number of unread emails in your inbox went from 23 to 5. What is the change in the number of unread emails?

21. Let x and z represent the values of two different quantities.
    a. If the value of x decreases from $x = 2$ to $x = -5$, what is the change in the value of x?
    b. If the value of x decreases from $x = 212$ to $x = 32$, what is the change in the value of x?
    c. If the value of z decreases from $z = 2.145$ to $z = 1.234$, what is the change in the value of z?

22. Let a and b represent the values of two different quantities.
    a. If the value of a decreases from $a = 3$ to $a = -4$, what is the change in the value of a?
    b. If the value of a increases from $a = -31$ to $a = 12.2$, what is the change in the value of a?
    c. If the value of b decreases from $b = -1.15$ to $b = -4.21$, what is the change in the value of b?

23. Fill in the following tables showing the appropriate changes in the value of the variable.

x	Δx
3	
6.3	
−2.6	

r	Δr
1	
3	
8.5	

t	Δt
−6	
−3.4	
2	

24. Fill in the following tables showing the appropriate changes in the value of the variable.

s	Δs
−12.43	
0.73	
−7.3	

y	Δy
2.85	
14.3	
−1.05	

p	Δp
3.834	
−2.3	
0	

25. This morning Tom went for a run. Let d represent the number of miles that Tom has run. What is the difference in meaning between $d = 11$ and $\Delta d = 11$?

26. Use the graph below to answer the following questions;
    a. Determine Δx and Δy from the point on the left to the point on the right. Illustrate the values of Δx and Δy on the graph below.
    b. Determine Δx and Δy from the point on the right to the point on the left. Illustrate the values of Δx and Δy on the graph below.

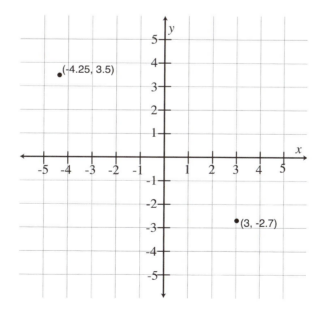

27. The number of calories Monica burns while running increases by 105 calories for every mile that Monica runs up to 15 miles. Let *n* represent the number of miles that Monica has run.
    a. Suppose the value of *n* increasess $n = 3$ to $n = 7.5$.
        i. What is the change in the value of *n*?
        ii. What does this change represent in the context of this situation?
        iii. What is the corresponding change in the number of calories Monica burns while running?
    b. How many calories will Monica burn for *any* 4.5 mile change in the number of miles Monica has run during a 15 mile run?
    c. For the following changes in the number of miles that Monica has run determine the corresponding change in the number of calories that Monica has burn.
        i. $\Delta n = 3.5$
        ii. $\Delta n = 0.41$
        iii. $\Delta n = 13.7$
        iv. $\Delta n = k$ for some constant $k \leq 15$ miles

28. A bucket full of water has a leak. The bucket loses 71 mL of water every 5 minutes. Let *t* represent the number of minutes since the bucket started leaking
    a. As the number of minutes since the bucket started leaking increases from 13 minutes to 29 minutes, what is the corresponding change in the volume of water in the bucket?
    b. By how much will the volume of water in the bucket change when the number of minutes since the bucket started leaking changes by 1 minute? Explain how you determined your answer.
    c. As the volume of water in the bucket decreases from 60 mL to 23 mL, what is the corresponding change in the number of minutes since the bucket started leaking?
    d. By how much will the number of minutes since the bucket started leaking change when the volume of water in the bucket decreases by 1 mL?

29. A group of Kansas University students were traveling from Lawrence, KS to Denver, CO for a weekend ski trip. On the way they stopped for a late dinner, then continued on to Denver driving through the night. They left the restaurant, located 112 miles from Lawrence, at 10:00pm and arrived at Denver, 565 miles from Lawrence, at 5:45 am. For the purpose of this problem, assume the car maintained a constant speed from the time they left the restaurant to the time they arrived in Denver.
    a. Explain what it means to say the car maintained a constant speed from the time it left the restaurant to the time it arrived in Denver. Make sure to explain the relationship it implies – do not say the car's speed does not change.
    b. At what constant speed did the car travel between the restaurant and Denver?
    c. The driver was listening to music to keep awake. Between 2:03 am and 2:55 am the driver listened to his favorite album.
        i. What was the change in the time elapsed while he listened to the album in minutes? In hours?
        ii. How far did the car travel while the driver listened to this album?
    d. As the Kansas University students traveled between two towns they noticed that their trip odometer reading changed from 234.6 miles from Lawrence to 302.4 miles from Lawrence.
        i. What was the change in distance from Lawrence between these two towns?
        ii. How much time elapsed while the car traveled between these two towns?
    e. Sketch a graph of the relationship between the students' distance from Lawrence (in miles) and the number of hours since the students left the restaurant. What does the slope of the graph convey about this situation?

30. Answer the following questions about the idea of constant rate of change.
    a. Joanne is purchasing fabric. The cost of the fabric she wants increases at a constant rate of $7.25 per yard. What does this mean for any change in the number of yards of fabric purchased?
    b. Between 1990 to 2003, the concentration of carbon monoxide in the atmosphere decreased at a constant rate of 0.248 parts per million per year. What does this mean for any change in the number of years since 1990?
    c. Between 1980 and 2004, the number of Medicare enrollees increased by 0.554 million people per year. What does this mean for any change in the number of years since 1980?

III. CONSTANT RATE OF CHANGE AND LINEAR FUNCTIONS (Text: S2, 3)

31. You are driving on the interstate with your cruise control on at a constant speed of 64 miles per hour. Use the number lines below to determine how long it will take to drive to the next rest sop that is 16 miles away (see textbook, page 20). Explain the thinking you used to determine your answer.

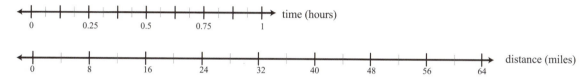

32. When an object is dropped, gravity pulls on the object and causes its speed to increase. The table below shows a certain object's speed at various moments during its fall. Does the object's speed (in feet per second) change at a constant rate with respect to the number of seconds since the object started falling? Explain your reasoning. If so, determine the value of the constant rate of change of the object's speed with respect to the number of seconds since the object started falling.

Number of seconds since the object started falling	The speed of the object (in feet per second)
0.15	4.83
0.4	12.88
0.52	16.744
0.98	31.556
1.26	40.572

33. The following table of values provides information about the distance of an airplane from Sky Harbor International Airport in terms of the number minutes since the plane took off. Does the distance of the airplane from Sky Harbor International Airport change at a constant rate with respect to the number of minutes since the plane took off? Explain your reasoning. If so, determine the value of the constant rate of change of the distance of the airplane from Sky Harbor International Airport in terms of the number of minutes since the plane took off.

Number of minutes since the airplane took off	Distance of the airplane from Sky Harbor International Airport (in miles)
3	16
5	32
9	64
11	92
18	170

34. Suppose that the quantities whose values are represented by x and y are related by a constant rate of change of y with respect to x.
    a. Given the information in the following table determine the value of m, the constant rate of change of y with respect to x.

x	y
-3.5	7.1
-1	-0.9
2	-10.5
6	-23.3
10	-36.1

    b. Given the information in the following graph determine the value of the constant rate of change of y with respect to x.

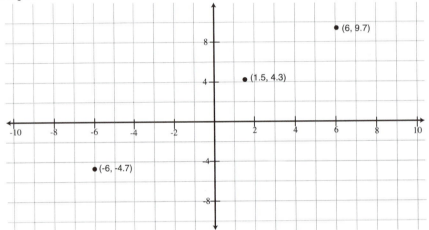

35. Given values of x and y in the tables below, which table(s) contain values that could define a linear relationship between two quantities? If the two quantities whose values are represented by x and y can be related by a linear relationship, what is the constant rate of change (the value of m)?

Table 1	
x	y
−2	−14
3	1
5	7
8	16
12	28

Table 2	
x	y
1	3
2	7
4	8
5	12
9	14

Table 3	
x	y
−5	9.5
−2	5
−1	3.5
3	−2.5
10	−13

36. For each situation below determine if both the changes in the two quantities and the two quantities are proportional.
    a. Between 2000 and 2005 the Burger Company's profit increased by $3,500 per year. In 2000 the Burger Company's profit was $52,000.
       i. Determine if the change in the number of years since 2000 is proportional to the change in the Burger Company's profit. Explain your reasoning.
       ii. Determine if the number of years since 2000 is proportional to the Burger Company's profit. Explain your reasoning.

(problem continues on next page)

b. You are planning a trip to Las Vegas and need to rent a car. After contacting the different car companies you choose to go with the company that charges a $25 rental fee and $0.05 per mile that the car is driven.
  i.   Determine if the change in the number of miles driven is proportional to the change in the total cost of the rental car. Explain your reasoning.
  ii. Determine if the number of miles driven is proportional the total cost of the rental car. Explain your reasoning.
c. When baking chocolate chip cookies you need 3 cups of flour per cup of sugar.
  i.   Determine if the change in the number cups of flour is proportional to the change in the number of cups of sugar. Explain your reasoning.
  ii. Determine if the number of cups of flour is proportional to the number of cups of sugar. Explain your reasoning.

37. Nick is considering joining a weight loss club that provides meals and support for people who want to lose weight. Based on an initial consultation with a weight loss advisor, Nick charted his potential weight loss based on the advisor's estimates of his expected weekly weight loss.

Number of weeks since joining	Nick's projected weight (pounds)
3	277
6	266.5
8	259.5
12	245.5
13	242

a. Are the quantities Nick's projected weight (in pounds) and time since joining (in weeks) proportional? Justify your answer.
b. Complete the following table showing the relative changes in the quantities change in time since joining (in weeks) and change in Nick's projected weight (in pounds).

Change in the number of weeks since joining	Number of weeks since joining	Nick's projected weight (in pounds)	Change in Nick's projected weight (pounds)
	3	277	
	6	266.5	
	8	259.5	
	12	245.5	
	13	242	

c. Do the quantities change in Nick's projected weight (in pounds) and change in time since joining (in weeks) appear to be proportional? Justify your answer.
d. Construct a graph showing the relationship between the quantities Nick's projected weight (in pounds) and time since joining (in weeks).
e. What does the slope of the graph represent in this context?

38. For the tables below,
    a. Determine if the quantities are proportional.
    b. Determine if the changes in the quantities are proportional.
    c. Determine if the relationship is linear. If the table represents a linear relationship, write a formula to represent how the quantities change together

Table 1	
x	y
2	1.42
−1.2	−0.852
5.1	3.621

Table 2	
a	b
0.7	1.3
−2	5.62
5.4	−2.885

Table 3	
r	s
2.4	0.76
0.3	3.595
−1.5	6.025

IV. CONSTANT RATE OF CHANGE & LINEARITY (Text: S3)

For Problems #39-40: Let r represent the possible values of one quantity and let p represent the possible values of another quantity.

39. Suppose r changes at a constant rate of 2 with respect to p.
    a. What does this mean for any change in p?
    b. If p changes by 6, how much does r change?
    c. If p changes by -3.1, how much does r change?

40. Suppose r changes at a constant rate of -1.3 with respect to p.
    a. What does this mean for any change in p?
    b. If p changes by 2, how much does r change?
    c. If p changes by -6.2, how much does r change?

41. Suppose the constant rate of change of y with respect to x is 0.17 and we know $y = 12.25$ when $x = 7.35$.
    a. What is the value of y when $x = 11.1$?
    b. What is the value of y when $x = -5.6$?
    c. What is the value of x when $y = 2.5$?
    d. What is the change in the value of y when the change in the value of x is 4.75?

42. Suppose the constant rate of change of y with respect to x is -11.1 and we know $y = -2.6$ when $x = -0.85$.
    a. What is the value of y when $x = 3.4$?
    b. What is the value of y when $x = -12.7$?
    c. What is the value of x when $y = 4.5$?
    d. What is the change in the value of y when the change in the value of x is -6.7?

43. A large spool is used to hold rope, which is wound around the spool. The more rope wound around the spool, the greater the combined weight of the spool and rope. The graph below shows that when 5 feet of rope is wound around the spool, the total weight of the spool and rope is 3.95 pounds. Note that the rope weighs 0.27 pounds per foot.

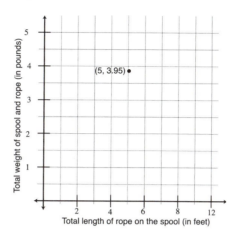

a. Suppose the number of feet of rope on the spool increases from the given point to 8.4 feet. What is the change in the number of feet of rope on the spool? Represent this change on the graph above.

b. By how much does the weight of the spool and rope change for the change in the amount of rope on the spool you found in part (a)? Represent this change on the graph above.

c. What is the total weight of the spool and rope when there are 8.4 feet of rope on the spool? Explain how you determined this.

d. What is the weight of the spool without any rope? Explain how you determined this value and represent your reasoning on the graph above.

44. When a bathtub made of cast iron and porcelain contains 60 gallons of water the total weight of the tub and water is approximately 875.7 pounds. You pull the plug and the water begins to drain. (Note that water weighs 8.345 pounds per gallon).

a. Describe the quantities in this situation. Which of these quantities are constant and which are changing?

b. Suppose that some water has drained from the tub and 47 gallons of water remain in the tub.
   i. What was the change in the number of gallons of water (recall the situation begins with 60 gallons of water in the tub)?
   ii. What is the corresponding change in the total weight of the tub and water?
   iii. What is the weight of the tub and water when there area 47 gallons of water in the tub?

c. Complete the following table of values.

Number of gallons of water remaining in the tub	Total weight of the tub and water (in pounds)
59	
40	
30	
20	
10	
8.5	

d. Suppose you and a friend can each lift about 150 pounds. Once empty, could you and your friend pick up and carry the bathtub out of the bathroom? Explain your reasoning.

45. John inserts a partially used battery into a portable electric fan. The percent of the battery's total charge changes at a constant rate of − 3.1% per minute since the fan was totally charged.

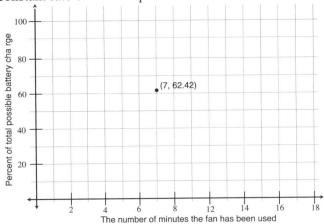

a. Represent "an increase of 10 minutes in the amount of time the fan is used" from the given reference point on the graph.
b. By how much will the percent of the total possible battery charge change when the fan is used for 10 minutes? Represent this on the graph.
c. What is the percent of the battery's total possible charge when the fan has been used for 17 minutes? Explain how you determined this value.
d. What is the vertical intercept of the graph of the function? Explain how you determined this value and represent your approach on the graph above. What does the vertical intercept represents in the context of this situation.

46. Consider the graph provided below and assume that the values of x and y change together at a constant rate of change.

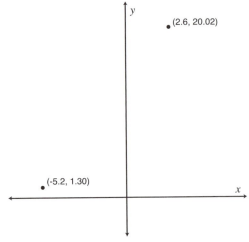

a. Determine the constant rate of change of y with respect to x.
b. Using the point (− 5.2, 1.30) as a reference point, what is the change in x from this point from to $x = 2$. Represent this change using the given axes.
c. What is the change in y that corresponds with the change in x found in part (b)?
d. What is the value of y when $x = 2$?
e. What is the vertical intercept of the function? Explain how you can find this value using the meaning of constant rate of change and the reference point (− 5.2, 1.30).

47. Consider the graph provided below and assume that the values of *x* and *y* change together at a constant rate of change.

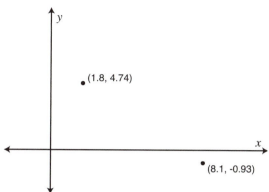

a. Determine the constant rate of change of *y* with respect to *x*.
b. Using the point (1.8, 4.74) as a reference point, what is the change in *x* from this point from to *x* = 3.6. Represent this change using the given axes.
c. What is the change in *y* that corresponds with the change in *x* found in part (b)?
d. What is the value of *y* when *x* = 3.6?
e. What is the vertical intercept of the function? Explain how you can find this value using the meaning of constant rate of change and the reference point (1.8, 4.74)

48. Two quantities (A and B) co-vary such that Quantity A changes at a constant rate with respect to Quantity B. Suppose the change in the value of Quantity A is always 6.5 times as large as the change in the value of Quantity B. Suppose also that the value of Quantity A is 5 when the value of Quantity B is 12.
a. What is the value of Quantity A when the value of Quantity B is 19?
b. What is the value of Quantity A when the value of Quantity B is – 1?
c. What is the value of Quantity B when the value of Quantity A is 7?

49. Two quantities (A and B) co-vary such that Quantity A changes at a constant rate with respect to Quantity B. Suppose the change in the value of Quantity A is always – 0.82 times as large as the change in the value of Quantity B. Suppose also that the value of Quantity A is −1.4 when the value of Quantity B is −2.5.
a. What is the value of Quantity A when the value of Quantity B is 3?
b. What is the value of Quantity A when the value of Quantity B is −7.7?
c. What is the value of Quantity B when the value of Quantity A is −6?

50. Given the values of *x* and *y* in the tables that follow, which table(s) contain values that could define a linear relationship between the two quantities? Explain your reasoning. For the table(s) that could represent a linear relationship, write a formula to define how the quantities change together.

Table 1	
x	*y*
1.25	3
3	6.5
5.5	11.5

Table 2	
x	*y*
3.25	10
3	9
5	4

Table 3	
x	*y*
3	10
6	20
7	70/3

51. Given the values of x and y in the tables that follow, which table(s) contain values that could define a linear relationship between the two quantities? Explain your reasoning. For the table(s) that could represent a linear relationship, write a formula to define how the quantities change together.

Table 1	
x	y
2	10.4
5	30.65
10	64.5

Table 2	
x	y
1.3	12.015
–4	14.4
9.7	8.235

Table 3	
x	y
–2	13.6
1	2.2
4.3	–10.34

52. Find a formula for each of the linear functions whose graphs are described below.
    a. The graph of the function with a constant rate of change of y with respect to x is 2.6 and passes through the point (–6, 4.2).
    b. The graph of the function with a constant rate of change of y with respect to x is $\frac{-4}{3}$ and passes through the point $\frac{9}{2}, 10$.
    c. The graph of the function passing through the points (–4, 4) and (3, –25).
    d. The graph of the function passing through the points $\frac{11}{2}, \frac{5}{3}$ and $\frac{3}{5}, \frac{-9}{7}$.

53. Find a formula for each of the linear functions whose graphs are described below.
    a. The graph of the function that passes through the point (2, –18.4) and the change in y is always –1.34 times as large as the change in x.
    b. The graph of the function with a constant rate of change of y with respect to x is $\frac{7}{9}$ and passes through the point $\frac{-3}{5}, \frac{2}{11}$.
    c. The graph of the function passing through the points (–2, 14) and (–12, –7.6).
    d. The graph of the function passing through the points $\frac{11}{9}, \frac{15}{7}$ and $\frac{-12}{4}, \frac{15}{7}$.

54. The graph of a certain linear function passes through the points (2,9) and (7, –11).
    a. What is the constant rate of change of y with respect to x (slope) for the function?
    b. From the point (2, 9) how much must x change to reach a value of $x = 0$?
    c. What is the corresponding change in the value of y for the change in x you found in part (b)?
    d. What is the value of y when $x = 0$?
    e. Write a formula to calculate the value of y for any value of x.

55. The graph of a certain linear function passes through the points (–7, –15) and (5, –7).
    a. What is the constant rate of change of y with respect to x (slope) for the function?
    b. From the point (–7, –15) how much must x change to reach a value of $x = 0$?
    c. What is the corresponding change in the value of y for the change in x you found in part (b)?
    d. What is the value of y when $x = 0$?
    e. Write a formula to calculate the value of y for any value of x.

56. Consider the formula $y = -12.13x + 7.14$. Suppose we want to find the value of y when $x = -1.15$. Explain how the formula determines the value of y using the meaning of constant rate of change. (You may sketch a graph or diagram if it helps you explain.)

57. The formula $a = 10 - 1.5t$ defines the remaining height (in inches) of a burning candle, a, in terms of the number of hours that the candle has been burning, t.
    a. What does 10 represent in the context of this situation?
    b. What does -1.5 represent in the context of this situation?
    c. What does $-1.5t$ represent in the context of this situation?
    d. Explain what the point $(t, a) = (2, 7)$ conveys in the context of this situation.
    e. What is the value of a when $t = 4.2$. Explain what this value represents in the context of this situation.

58. Determine a formula that defines the linear functions whose graphs are described below.
    a. The graph of the function conveys that the constant rate of change of y with respect to x is $\frac{2}{3}$ and the vertical intercept is 5.
    b. The graph of the function conveys that the constant rate of change of y with respect to x is 5 and the graph crosses the vertical axis at $(0, -2)$
    c. The graph of the function conveys that the constant rate of change of y with respect to x is $\frac{-6}{7}$ and the vertical intercept of $\frac{-1}{10}$.

59. Write the formula that defines the linear relationship given in each of the following graphs.
    a.

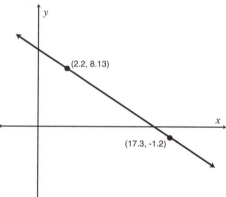

    b.

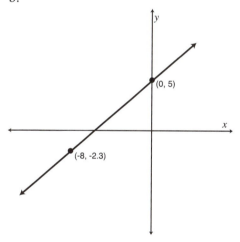

60. Use the graph below to answer the following questions.

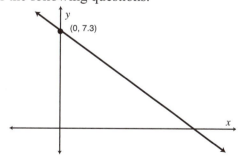

    a. The graph conveys that the constant rate of change of y with respect to x is -4.1. What is the value of y when $x = 5.15$? Explain how to find this value using the meaning of constant rate of change. Draw a diagram to help represent your reasoning.
    b. What is the value of y when $x = -3.81$? Explain how to find use the idea of constant rate of change to find the value of y.

61. A tortoise and hare are competing in a race around a 1600-meter track. The arrogant hare decides to let the tortoise have a 630-meter head start. When the start gun is fired, the hare begins running at a constant speed of 8.5 meters per second and the tortoise begins crawling at constant speed of 6 meters per second.
    a. What quantities are changing in this situation? What quantities are not changing? (Be sure to include the units of each quantity.)
    b. Define a formula to determine the distance of the tortoise from the starting line in terms of the amount of time since the start gun was fired.
    c. Define a formula to determine the distance of the hare from the starting line in terms of the amount of time since the start gun was fired.
    d. The tortoise traveled 170 meters as he moved from his starting position to a curve on the track. If possible, find the following:
        i. The amount of time that it took the tortoise to travel the 170-meter distance.
        ii. The amount of time that it took the tortoise to travel the next 80 meters on the track.
    e. Now we will consider how the distance between the tortoise and the hare changes throughout the race.
        i. Explain how the distance between the tortoise and hare changes as the number of seconds since the start gun was fired increases.
        ii. Define a formula that relates the distance between the tortoise and the hare with the number of seconds since the start gun was fired.
        iii. Is the relationship you defined in part (ii) linear? Explain.
        iv. Who finishes the race first, the tortoise or the hare? Explain.

62. Lisa and Sarah decided to meet at a park bench near both of their homes. Lisa lives 1850 feet due west of the park bench and Sarah lives 1430 feet due east of the park bench. Sarah left her house at 7:00 pm and traveled at a constant speed of 315 feet per minute towards the bench. Lisa left her house at the same time traveling a constant speed of 325 feet per minute towards the bench.
    a. Illustrate this situation with a drawing, labeling the constant and varying quantities.
    b. Define a formula to relate Lisa's distance (in feet) from the park bench in terms of the number of minutes that have passed since 7:00 pm. (Define relevant variables.)
    c. Define a formula to relate Sarah's distance (in feet) from the park bench in terms of the number of minutes that have passed since 7:00 pm. (Define relevant variables.)
    d. Who will reach the park bench first? Explain your reasoning.
    e. Now we will consider how the distance between Lisa and Sarah changes as the number of minutes since 7:00 pm increases.
        i. Explain how the distance (in feet) between Lisa and Sarah changes as the amount of minutes since 7:00pm increases.
        ii. Define a formula that relates the distance between Lisa and Sarah in terms of the number of minutes since 7:00pm.

63. John and Susan leave a neighborhood restaurant after having dinner. They each walk to their respective homes. John's home is 3120 feet due north of the restaurant and Susan's home is 2018 feet due south of the restaurant. John leaves at 7:28 pm traveling at a constant speed of 334 feet per minute and Susan leaves at 7:30 pm traveling at a constant speed of 219 feet per minute.
    a. Illustrate this situation with a drawing and define relevant variables.
    b. Define a formula to relate John's distance from the restaurant in terms of the number of minutes that have elapsed since 7:30 pm.
    c. Define a formula to relate Susan's distance from the restaurant in terms of the number of minutes that have passed since 7:30 pm.

(problem continues on next page)

    d. Will John or Susan arrive home first? Explain your reasoning.
    e. Now we will consider how the distance between John and Susan changes as the number of minutes that have passed since 7:30 pm increases.
       i. Explain how the distance (in feet) between John and Susan changes as the number of minutes that have passed since 7:30 pm increases.
      ii. Define a formula that relates the distance (in feet) between John and Susan in terms of the number of minutes that have passed since 7:30pm.

64. Lucia rented a car for $211 per week (with unlimited driving miles) to use during her spring break vacation. She must also pay for gasoline which costs $3.65 per gallon. The gas mileage of the car is 30 miles per gallon (mpg) on average.
    a. What quantities are changing in this situation? What quantities are not changing?
    b. i. If Lucia traveled 100 miles, how many gallons of gasoline did she use?
      ii. If Lucia traveled 200 miles, how many gallons of gasoline did she use?
      iii. If Lucia traveled 500 miles, how many gallons of gasoline did she use?
    c. Define a formula to determine the number of gallons of gasoline n used in terms of the number of miles driven x.
    d. Define a formula to determine the cost of gasoline c in terms of the number of gallons of gasoline used.
    e. Define a formula to determine the total cost T of driving the rental car x miles, including the one-week rental cost of $211.
    f. How much does the total rental cost increase for each 100 miles the car is driven during the rental period. Explain your reasoning.
    g. How much does the total rental cost increase for each mile driven? Explain the thinking you used to arrive at your answer.

65. Sketch a graph of the following relationships.
    a. $3.25 = y$
    b. $6 = x$
    c. $y = 2x + 4.3$
    d. $y = -\frac{1}{2}x + 3$

66. On the same axes sketch a graph of $y = -4x - 2$ and $y = \frac{1}{4}x + 3$. What do you notice about the graphs of two linear relationships?

67. Simplify the following expressions.
    a. $2x + 7 - 3x - 2$
    b. $\frac{3}{7}x - (-1 + \frac{2}{3}x)$
    c. $3x - (-7x) + 4 - 2.2$
    d. $2(x - 4) + 3x - (\frac{9}{8}x - 7)$

V. Exploring Average Speed (Test: S4)

68. The following graph represents the distance-time relationship for Kevin and Carrie as they cycled on a road from mile marker 225 to mile marker 230.

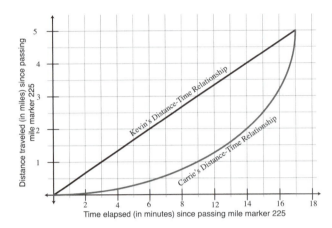

   a. How does the distance traveled and time elapsed compare for Carrie and Kevin as they traveled from mile marker 225 to mile marker 230?

   b. How do Carrie and Kevin's speeds compare as they travel from mile marker 225 to mile marker 230?

   c. How do Carrie and Kevin's average speeds compare over the time interval as they traveled from mile marker 225 to mile marker 230?

   d. Do Carrie and Kevin collide on the course 17 minutes after the passed mile marker 225?

69. When running a marathon you heard the timer call out 12 minutes as you passed mile-marker 2.

   a. What quantities could you measure to determine your speed as you ran the race? Define variables to represent the quantities' values and state the units you will use to measure the value of each of these quantities.

   b. As you passed mile-marker 5 you heard the timer call out 33 minutes. What was your average speed from mile 2 to mile 5?

   c. Assume that you continued running at the same constant speed as computed in (b) above. How much distance did you cover as your time spent running increased from 35 minutes after the start of the race to 40 minutes after the start of the race?

   d. If you passed mile marker 5 at 33 minutes, what average speed do you need to run for the remainder of the race to meet your goal to complete the 26.2-mile marathon in 175 minutes?

   e. What is the meaning of average speed in this context?

70. When running a road race you heard the timer call out 8 minutes as you passed the first mile-marker in the race.

   a. What quantities will you measure to determine your speed as you travel? Define variables to represent the quantities' values and state the units you will use to measure the value of each of these quantities.

   b. As you passed mile-marker 6 you heard the timer call out 52 minutes. What was your average speed from mile-marker 1 to 6?

   c. After mile-marker 6 you slowed down and ran at a constant rate of 10 minutes per mile between mile-marker 6 and mile-marker 10, how many minutes did it take you to travel from mile marker 6 to 9?

71. Marcos traveled in his car from Phoenix to Flagstaff, a distance of 155 miles.
    a. Determine the amount of time required for Marcos to travel from Phoenix to Flagstaff if his average speed for the trip was 68 miles per hour.
    b. Construct a possible distance-time graph of Marcos's trip from Phoenix to Flagstaff. Be sure to label your axes.
    c. On the same axes, construct a graph that represents the distance-time graph that represents another car traveling at a constant speed for the entire trip.

72. On a trip from Tucson to Phoenix via Interstate 10, you used your cruise control to travel at a constant speed for the entire trip. Since your speedometer was broken, you decided to use your watch and the mile markers to determine your speed. At mile marker 219 you noticed that the time on your digital watch just advanced to 9:22 am. At mile marker 197 your digital watch advanced to 9:46 am.
    a. Compute the constant speed at which you traveled over the time period from 9:22am to 9:46 am.
    b. As you were passing mile marker 219 you also passed a truck. The same truck sped by you exactly at mile marker 197.
        i.   Construct a distance-time graph of your car. On the same graph, construct one possible distance-time graph for the truck. Be sure to label the axes.
        ii. Compare the speed of the truck to the speed of the car between 9:22am and 9:45am.
        iii. Compare the distance that your car traveled over this part of the trip with the distance that the truck traveled over this same part of the trip. Compare the time that it took the truck to travel this distance with the time that it took your car to travel this distance. What do you notice?
        iv. Why are the average speed of the car and the average speed of the truck the same?
        v.   Phoenix is another 53 miles past mile marker 197. Assuming you continued at the constant speed, at what time should you arrive in Phoenix?

73. The distance d (measured in a number of feet) between Silvia and her house is modeled by the formula , $d = t^2 + 3t + 1$ where t represents the number of seconds since Silvia started walking.
    a. Find Silvia's average speed for the time period from $t = 2$ to $t = 7$ seconds.
    b. What was Silvia's change in distance as the time since Silvia started walking increased from 2 to 7 seconds?
    c. i.   Construct a graph that gives Silvia's distance from her house (in feet) in terms of the number of seconds since she started walking. Be sure to label your axes.
        ii. Illustrate (with a line segment on the graph you constructed in part (i)) Silvia's change in distance during the time period from $t = 4$ and $t = 5$ seconds since she started walking.
        iii. Illustrate (with a line segment on the graph you constructed in part (i)) Silvia's change in distance during the time period from $t = 5$ and $t = 6$ seconds since she started walking.

74. Bob's distance, d, north of Mrs. Bess's restaurant (in feet) is given by the formula $d = 2t^2 - 7$ where t represents the number of seconds since Bob began driving.
    a. Determine the value of d when $t = 1$. What does a negative value for d represent in the context of this problem?
    b. Find the average speed of Bob's car for the time period from $t = 3$ to $t = 5$.
    c. As the number of seconds since Bob began driving increased from 1.5 to 2 seconds, by how much did Bob's distance north of Mrs. Bess's restaurant change?
    d. As the number of seconds since Bob began driving increased from 2 to 2.5 seconds, by how much did Bob's distance north of Mrs. Bess's restaurant change?
    e. i.   How much time did it take Bob to travel from 20 to 30 feet north of the restaurant?
        ii. How much time did it take Bob to travel from 30 to 40 feet north of the restaurant?
        iii. How much time did it take Bob to travel from 40 to 50 feet north of the restaurant?

75. The graph that follows represents the speeds of two cars (car A and car B) in terms of the elapsed time in seconds since being at a rest stop. Car A is traveling at a constant speed of 65 miles per hour. As car A passes the rest stop car B pulls out beside car A and they both continue traveling down the highway.

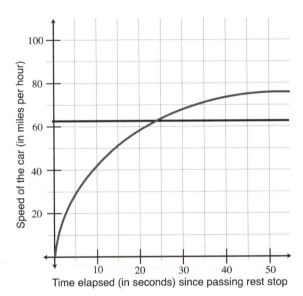

a. Which graph represents car A's speed and which graph represents car B's speed? Explain.
b. Which car is further down the road 20 seconds after being at the rest stop? Explain.
c. Explain the meaning of the intersection point.
d. What is the relationship between the positions of car A and car B 25 seconds after being at the rest stop?

Instructions for Problems 76-85: Let *d* be the distance of a car (in feet) from mile marker 420 on a country road and let *t* be the time elapsed (in seconds) since the car passed mile marker 420. The formulas below represent various ways these quantities might be related. For each of the following:
 i. Determine the average speed of the car using the given formula and the specified time interval.
 ii. Explain the meaning of average speed for the given situation.

76. $d = t^2$ from $t = 5$ to $t = 30$.

77. $d = -3(-19t - 1)$ from $t = 3$ to $t = 9$.

78. $d = 5(12t + 1) + 3t$ from $t = 0.5$ to $t = 3.75$.

79. $d = \frac{10t(t+5)-14}{2}$ from $t = 0$ to $t = 5$.

80. $d = \frac{1}{3}\left(9t^2 + 155t - (11t - 6)\right)$ from $t = 2$ to $t = 4$.

81. $d = (2t + 7)(3t - 2)$ from $t = 2$ to $t = 2.75$.

82. $d = \left(\frac{1}{3}t + 60\right)\left(t + \frac{1}{2}\right)$ from $t = 1$ to $t = 4$.

83. $d = (t+6)(t+3) + 7t - 20 + 11t - \frac{7}{8}t^2$ from $t = 30$ to $t = 35$.

84. $d = \frac{1}{8}t \; 3t^2 + 1.5t \; +16t - 3$ from $t = 2$ to $t = 4$.

85. $d = \dfrac{t \; \frac{1}{2}t^2 + 15 \; + 3t \; \frac{1}{10}t^2 + \frac{4}{3t}}{5}$ from $t = 5$ to $t = 9$.

VI. THE DISTANCE FORMULA (Text: S5)

86. Use the distance formula to find the distance between the points on the graph below.

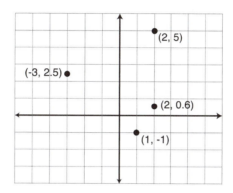

87. Suppose that three corners of a rectangle are located at the points (–2, –1), (7, 3.5) and (–2, 3.5) on the coordinate axes. What is the point where the fourth corner of the rectangle is located? Justify your answer.

88. Suppose the endpoints of one side of a rotated squared are located at (–5, –7) and (1, –1) on the coordinate axes. Find the dimensions and area of the square.

89. Suppose a circle is centered at the origin (0, 0) and has radius of length 4.
    a. For which points on the circle is the x-coordinate equal to 1?
    b. For which points on the circle is the y-coordinate equal to 3?
    c. What is the equation for the graph of this circle?

90. Suppose a circle is centered at the origin (–2, 4) and has radius of length 6.
    a. For which points on the circle is the x-coordinate equal to 0?
    b. For which points on the circle is the y-coordinate equal to 4?
    c. What is the equation for the graph of this circle?

91. The center of a circle is located at the point (1, 1). The point (–5, 2) is located on the circle.
    a. What is the radius of the circle?
    b. What is the equation for the graph of this circle?

92. The center of a circle is located at the point (4, 6). The point (12, 9) is located on the circle.
    a. What is the radius of the circle?
    b. What is the equation for the graph of this circle?

93. The center of a circle is located at the point (–5, 0). The point (–9, –3) is located on the circle.
    a. What is the radius of the circle?
    b. What is the equation for the graph of this circle?

VI. ABSOLUTE VALUE (Text: S6)

94. A contractor is digging a hole that needs to be 36" deep. He knows that his measurement is no farther than 1.5 inches different from the actual depth of the hole.
    a. Indicate on a number line the depths that the hole could be.
    b. Use algebraic symbols to represent values of x (hole depth values) that are less than or equal to 1.5 inches away from 36 inches (the desired depth of the hole).
    c. Use algebraic symbols to represent the difference between varying values of the depth of the hole, x, and 36 inches (the ideal depth of the hole), sometimes called the margin of error.
    d. Use symbols to represent values of the margin of error that are less than or equal to 1.5 inches.

95. The ideal weight of a bag of cookies is 541 grams. The actual weight may vary from the ideal by no more than 3 grams for the bag of cookies to be sold.
    a. Represent the possible weights of the bag of cookies on a number line.
    b. Use algebraic symbols to represent values of w (the weight of the bag of cookies) that are less than or equal to 3 grams away from 541 (the ideal weight of the bag of cookies).
    c. Use algebraic symbols to represent the difference between varying values of the weight of the bag of cookies, w, and 541 grams (the ideal weight of a bag of cookies), called the margin of error.
    d. Use symbols to represent values of the margin of error that are less than or equal to 3 grams.

96. a. Represent all numbers whose distance from 3 is less than 4 on a number line.
    b. What is the meaning of $-4 < x - 3 < 4$

97. a. Represent all numbers whose distance from –5 is less than 2 on a number line.
    b. What is the meaning of $-2 < x - \ -5 \ < 2$

98. Describe the values of x that are being described by the given inequalities.
    a. $-4 < x - 2 < 4$
    b. $-1 < x - \ -3.4 \ < 1$
    c. $x - 2.7 > 5$ and $x - 2.7 < -5$

99. Use absolute value notation to represent the following.
    a. All numbers x whose distance from 7 is less than 4.
    b. All numbers x whose distance from –2 is less than 3.4
    c. All numbers x whose distance from –5 is greater than 5.2

100. Illustrate the solutions to the given absolute value equations on a number line, then state the solutions algebraically.
    a. $|x| = 3.5$
    b. $|x| < 3.5$
    c. $|x| > 3.5$

101. Represent the solution set of the following absolute value inequalities by:
    i.   Describing the solution in words
    ii. Illustrating the solutions on a number line
    iii. Writing an inequality (with no absolute values).

   a. $|x-2|<4$

   b. $|x+7.2|<3$

   c. $|x-14|>5$

   d. $|x+7.3|>2.7$

102. Solve each of the following equations for x. Check your answers and show your work.

   a. $|x|=7$

   b. $|x-5|=3$

   c. $|4x-7|=15$

   d. $|19.25x-17.3|=-98.2$

   e. $|11x+22|=44$

   f. $|2x-2|=x+3$

   g. $|8x-9|=|2x+3|$

   h. $|9x-12|=|4x-5|$

*1. Consider what is involved in building a box (without a top) from an 8.5" by 11" sheet of paper by cutting squares from each corner and folding up the sides.

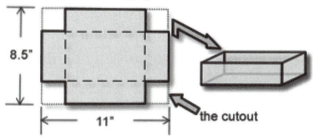

8.5"

11"

the cutout

a. To understand how the quantities in the situation are related it is important to first model the situation by:
   i. Cutting four equal-sized squares from the corners of an 8.5 by 11 inch sheet of paper
   ii. Folding up the sides and taping them together at the edges

b. Do the cutouts have to be square? Explain.

c. What quantities in this situation vary? What quantities in the situation are constant (do not vary)?

d. Describe how the configuration of the box changes as the length of the side of the square cutout varies.

e. Using the "Volume w/ Cubes" animation, describe how the volume of the box varies as the length of the side of the cutout varies from 0 to 4.25 inches.
   As the V will increase until it reaches max v then it will decrease

f. Based on your response to part (e), sketch a graph of the volume of the box (in cubic inches) with respect to the length of the side of the square cutout (in inches).

g. Use a ruler to measure the length of the side of your square cutout (measured in inches), where the length of the base is the side that was originally 11 inches long and the width of the base is the side that was originally 8.5 inches long.

   Cutout length: _____

   Box's height: _____

   Length of box's base: _____

   Width of the box's base: _____

   Volume of the box: _____

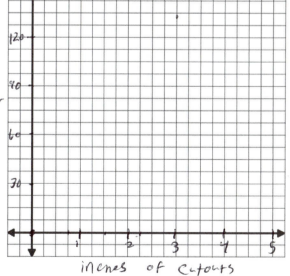

inches of cutouts

*2. Let x represent the varying length of the side of the square cutout in inches. Let w represent the varying width of the box's base in inches. Let l represent the varying length of the box's base in inches. Let V represent the varying volume of the box in cubic inches.

a. Complete the following table.

x	w	l	V
0	8.5		
1	6.5	9	58.5
4.25			
5			
		15	
		2	

b. Explain how the length of the base of the box is related to the length of the side of the square cutout and how the width of the base of the box is related to the length of the side of the square cutout.

$$L - 2x$$
$$W - 2x$$

c. Define a formula to determine the width of the box in terms of the length of the side of the square cutout. Be sure to define your variables.

$$W = 8.5 - 2x$$

d. Define a formula to determine the length of the box in terms of the length of the side of the square cutout. Be sure to define your variables.

$$L = 11 - 2x$$

e. Define a formula to determine the volume of the box in terms of the length of the side of the square cutout.

$$f(x) = (x)(11 - 2x)(8.5 - 2x)$$

f. Use your formula from part (e) to represent the volume of the box when x, the length of the side of the cutout, is 0.5 inches.

$$(.5)\left(11 - \cancel{2(x)}\right)(8.5 - 1) = 37.5$$
$$(.5)(10) \quad (7.5)$$

g. Use your formula from part (e) to represent the volume of the box when x, the length of the side of the cutout, is 3 inches.

$$(3)(11 - 6)(8.5 - 6) = 37.5$$

3. Represent the volume of the box for cutout lengths of 1.5, 2.7, 3.8, and 4.2 inches using your formula.

   a. When $x = $ **1.5** inches, $V = $ **66** cubic inches

   b. When $x = $ **2.7** inches, $V = $ **46.8** cubic inches

   c. When $x = $ **3.8** inches, $V = $ **11.628** cubic inches

   d. When $x = $ **4.2** inches, $V = $ **1.092** cubic inches

---

In math we frequently want to refer to the value of a quantity without having to calculate it. We also want to refer to the rule (or formula) for calculating a quantity's value without having to write the rule repeatedly as you did above. The convention we use is called **function notation,** and we will elaborate on the components and uses of function notation below.

To help with clarity and ease of communication, we will define a function named *f* to determine the volume of the box in cubic inches in terms of *x*, the length of the side of the square cutout of that box in inches. The rule for this function is the same as the formula you have just developed in previous exercises, but now we will instead represent the varying values of the volume of the box in cubic inches *V* using function notation, *f(x)*, so that we can more clearly refer to the value of *x* that corresponds to a particular box volume. (It is important to realize that $V = f(x)$ as *x* and *V* vary together.)

Function Definition

$$f\ (\ x\ ) = (x)(11 - 2x)(8.5 - 2x)$$

The convention for using function notation is that you write the name of the function, the variable that the rule acts on or takes as input, and then the rule that defines the function. We can use the phrases "name of rule" and "name of function" interchangeably. The values that we put into the rule are called *independent (or input) values*. The number that results from applying the rule to an *independent (or input) value* is called *a dependent (or output) value*. The symbol *f(x)* is an example of *representing the function's dependent values* for varying values of the independent quantity, *x*. The act of using function notation to represent a relationship between two quantities' values is called *defining a function*.

Given that the function named *f* is defined as above, then
  $f(0.5)$ represents the box's volume when the length of the side of the cutout is 0.5 inches. Also, $f(0.5)$ is the *output* of *f* when given 0.5 as *input*.
  $f(x)$ represents the box's volume when the length of the side of the cutout is *x* inches. Notice that *x* can vary as the cutout length varies. As *x* varies, $f(x)$ varies too.

We will use function notation repeatedly throughout these modules, so it is important that you are able to read and write relationships between two quantities' values with this notation.

---

   e. What is the smallest possible value for *x*, the length of the side of the square cutout? **0**

   f. What is the largest possible value for *x*, the length of the side of the square cutout? **4.25**

   g. We define the ***domain*** as the set of all possible independent values for our function. Relative to the box folding context, what is the domain of the function, $V = f(x) = (x)(11 - 2x)(8.5 - 2x)$?

   $$d = [0, 4.25]$$

4. a. View the "Graphs" animation (or use your graphing calculator) to create a graph that represents the volume of the box V (measured in cubic inches) in terms of the length of the side of the square cutout x (measured in inches). (When determining the window setting on your calculator, consider the possible values of x and the possible values of V.) Construct the graph and label two points on the graph. State what each of these points conveys about the box.

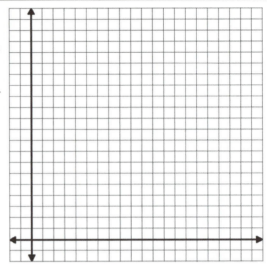

Point 1:

Point 2:

 b. Identify the point on the graph that corresponds to the dimensions of your box.

 c. As x (the length of the side of the cutout) increases from 0.5 to 0.75 inches, how does the volume of the box change?

 d. As x (the length of the side of the cutout) increases from 2.1 to 2.7 inches, how does the volume of the box change?

 e. As x (the length of the side of the cutout) increases from 1 to 3 inches, how does the volume of the box change?

 f. Indicate on the graph, a change of cutout length from 2 inches to 3 inches. Indicate on the graph the length that corresponds to the amount the volume changes when the cutout length increases from 2 inches to 3 inches.

 g. Estimate the interval(s) of values for the length of the side of the cutout x for which the volume of the box decreases.

*5. Using a graphing calculator, determine the following:
 a. An approximate value for the maximum value of the box.

 b. The length of the side of the square cutout when the box has maximum volume.

 c. The length of the side of the cutout when the volume of the box is 25 cubic inches.

*6. Let x represent the length (in inches) of a side of a square and let A represent the area of the square in square inches, so that $A = x^2$. Expand each expression if possible and describe what the expression represents.

    a. $3x^2$

    b. $(3x)^2$

    c. $(x+3)^2$

    d. $\dfrac{x}{4}$

    e. $\left(\dfrac{x}{4}\right)^2$

    f. $x^2 - \left(\dfrac{1}{3}x\right)^2$

7. Simplify the following expressions.

    a. $-3x+7-(4-x)$ b. $2.5x-3(-5x-10)$ c. $\dfrac{3}{4}x+\dfrac{1}{2}x\cdot x\cdot x-5\dfrac{1}{4}x$

8. a. Simplify $5(3x)^2 - 7(x)^2$ b. Simplify $\left(\dfrac{x}{5}\right)^2 - 3(-2x+5)$

9. Evaluate each of the following:

    a. $f(7)$ when $f(x)=\dfrac{x^2+(-2+x)}{7x-8}$.

    *b. $f(x+2)$ when $f(x)=2x^2+8x-12$.

In the previous investigation you related values of two quantities by writing formulas to describe how values of one quantity are related to values of another quantity. If the two quantities are related in such a way that each input to the formula generates exactly one output value we can say that the formula defines a function.

A **function** consists of three parts:
1. *Domain*: The values the independent quantity (also called *input* quantity) may assume.
2. *Range*: The values the dependent quantity (also called *output* quantity) may assume.
3. *Rule*: The rule that assigns to each value of the independent (input) quantity *exactly one* value of the dependent (output) quantity.

It is noteworthy that a rule of a function can be expressed using any of: i) a worded description; ii) an algebraic expression; iii) a graph; iv) a table of values

Terminology: We say *"p is a function of s"* meaning $p(s)$ represents the values of the dependent quantity and s represents the values of the independent quantity.

Equivalent phrases: Define *"p in terms of s"*; Define *"p with respect to s"*
(Recall that we think of variables such as s as representing the varying values of a quantity.)

1. Recall the box problem from Module 3 Investigation 1.
   a. Consider the function that defines the volume of the box in terms of the length of the side of the square cutout.
      i. What is the independent (input) quantity?

      ii. What is the dependent (output) quantity?

      iii. What is the rule of the function?

      iv. Is the relationship a function? Explain.

   b. Consider the length of the side of the square cutout in terms of the volume of the box.
      i. What is the independent quantity?

      ii. What is the dependent quantity?

      iii. Is the relationship a function? Explain.

2. Billy is walking from the front door of his house to his bus stop, which is 960 feet away from his front door. As Billy walks out his front door he walks in a straight path towards his bus stop at a constant rate of 7.5 feet per second.
   a. Illustrate the situation with a diagram and define variables to represent the values of the relevant varying quantities. (Label the variables on your picture.)

b. Define a function *f* to determine Billy's distance from his bus stop in terms of the number of seconds he has been walking.

c. What is the independent quantity and what is the domain of *f* (the values the independent quantity can take on)?

d. What is the dependent quantity and what is the range of *f* (the values the dependent quantity can take on)?

e. What is the meaning of *f*(0), *f*(6.5), *f*(100), and *f*(128) ?

f. Use function notation to express the following:
   i. Billy's distance from the bus stop after he has walked 23.6 seconds.

   ii. How much Billy's distance from the bus stop changes, as the number of seconds he has been walking since he left his front door increases from 35 seconds to 48 seconds.

g. Assuming *t* represents the number of seconds since Billy left his front door:
   i. Solve $f(t) = 150$ for *t*

   ii. What is the meaning of the statement "solve $f(t) = 150$ for *t*" ?

3. As Pat was driving his hybrid cruiser across Kansas with his cruise control on, his gas gauge broke. At the moment the gauge broke, he had 15 gallons of gas in the car's gas tank and his gas mileage was 42 miles per gallon (assume that he maintains this gas mileage by leaving cruise control on). Pat needs to keep track of how much gas is left in his gas tank.
   a. How many gallons does Pat have left after he has driven 84 miles? 150 miles?

   b. Define a function *f* to determine the number of gallons left in the tank $f(x)$ in terms of the number of miles driven *x*.

   c. What is the domain of *f*?

   d. What is the range of *f*?

   e. What does *f*(100) represent in the context of this problem?

   f. Explain what $\dfrac{x}{42}$ represents in the context of this situation.

   g. Explain what the expression $15 - \dfrac{x}{42}$ represents in the context of this situation.

   h. What are the maximum and minimum values that $f(x)$ can assume in the context of this situation? Explain. (Note: The maximum value that $f(x)$ can be is also called the maximum value of the function *f*.)

i.   Construct a graph of *f* on the given axes. Explain what the graph of *f* conveys about how the number of miles Pat has driven since leaving home *x* and the number of gallons of gas left in Pat's tank *f(x)* change together.

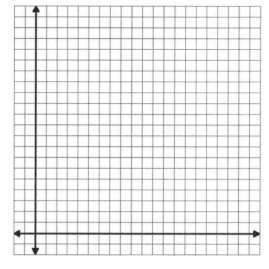

j.   What does the point (0, 15) represent in the context of this situation?

The Vertical Intercept of a Function

Given the graph of a function *f* where $f(0) = b$, the point $(0, b)$ is the point where the graph of *f* crosses the *vertical* axis and is known as ***the vertical intercept of f***.

---

k.   Since the vertical intercept of a function *f* occurs where $x = 0$, what general method can you use to determine the vertical intercept of the graph of a function *f*?

l.   What is the value of *x* when $f(x) = 0$? Show work and explain your approach. What point on the graph of *f* corresponds to where $f(x) = 0$? What does this point convey about the situation?

The Horizontal Intercept of a Function

Given the graph of a function *f* and the point $(a, 0)$ that is on the graph of *f*, this point is where the graph of *f* crosses the horizontal-axis. We can determine the value of *a* by substituting 0 in for *f(x)* and solving for *x*.

---

m.   Since the horizontal intercept of a function *f* occurs where $f(x) = 0$, what general method can you use to determine the horizontal intercept of the graph of a function *f*?

n.   True or False: $f(20) < f(50)$. Explain.

o.   What does $f(200) - f(30)$ represent in the context of this situation?

---

4. For this graphical representation
   a. Determine if y is a function of x. Justify your answer using the definition of function—each input value of a function is assigned to exactly one output value.

   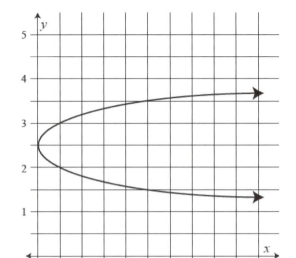

   b. Determine if x is a function of y. Justify your answer.

5. For this graphical representation
   a. Determine if y is a function of x. Justify your answer using the definition of function—each input value of a function is assigned to exactly one output value.

   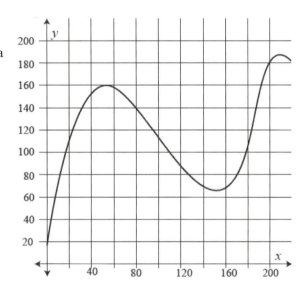

   b. Determine if x is a function of y. Justify your answer.

Recall that the *domain* is the set of all possible values of the independent quantity for which the function is defined.

6. Without using a graphing calculator determine the domain and range of the following functions.

   *a. $f(x) = \sqrt{x-4}$

   *b. $g(x) = \frac{1}{x^2-9}$

   *c. $h(x) = x^2 + 2x - 5$

   d. $k(x) = \frac{\sqrt{x-2}}{x-9}$

   e. $p(x) = \frac{x}{9}$

   f. $s(x) = \frac{9}{x}$

   g. $b(x) = \frac{x}{x-5}$

   h. $m(x) = \sqrt{2-3x}$

1. A hose is used to fill an empty wading pool. The graph shows volume (in gallons) in the pool as a function of time (in minutes) since the pool started filling.

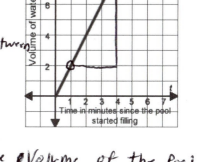

a. Define a function g that expresses the volume (or number of gallons) in the pool as a function of the time t in minutes since the pool started filling.

$$g(t) = 2t$$

b. Evaluate $g(4) - g(1)$ and describe what the answer represents in the context of this situation. The change in V in between 1 second and 4 seconds

c. Represent $g(4) - g(1)$ on the graph.

d. True or False: For any increase of 3 minutes since the pool started filling, the number of gallons in the pool increases by 3 gallons. Explain. False. In 3 minutes the Volume of the Pool should raise by 6 gallons

e. Describe the meaning of $g(4) + 100$ in the context of this situation.
The Pool was already had 100 g in it and then you added the hose and added 2g per minutes for 4 minuts leaving 108 g in the Pool.

f. Define a function h that expresses the time t (in minutes) since the pool started filling as a function of the number of gallons of water that are in the pool. (Let v represent the number of gallons of water in the pool after it has been filling for t minutes.)

$$h(v) = \frac{v}{2}$$

g. How are the functions g and h related? What is the independent quantity in the function g? What is the independent quantity in the function h?
Its the same function but the axis are fliped, in function g, time was the indipendent variable and in function h V is the independent variable.

2. Recall that the formula for determining the area of a circle is $A = \pi r^2$ where A is the area of the circle measured in square inches and r is the radius of the circle measured in inches. Let g be the name of the function that takes the radius length of a circle as an input and outputs the associated area of the circle.

a. Explain in your own words what $g(r)$ represents.

b. Use function notation to represent (NOT CALCULATE) the areas of three different circles whose radii are 3.5 inches, 18.2 inches, and 26.92 inches.

c. Interpret what $g(4.9) = 75.43$ means in the context of the problem.

© 2015 Carlson, Oehrtman, and Moore

d. What does it mean to solve $g(r) = 141.026$ for r. (Don't explain *how* to solve for r, instead explain what the solution would represent.)

e. Determine the value of r such that $g(r) = 141.026$.

*3. Suppose $P = h(t)$ represents the population of the city of Ames, Iowa t years after 1990.
  a. Using function notation, represent the population of Ames k years after 1990.

$$P = h(k) \bullet$$

  b. Describe the meaning of $h(k + 3)$.

$h(k+3)$ is the output of the function, so the pop. 3 years after 1990

  c. Using function notation, represent a population that is 765 more than the population of Ames in the year 1996.

~~$h(k)$~~ $h(k+6) =$

$$h(k+6) + 765$$

  d. Using function notation, represent the change in population of Ames from m years after 1990 to $m+7$ years after 1990.

$$h(m+7) - h(m)$$

  e. What does the expression $\dfrac{h(m + 7) - h(m)}{(m + 7) - (m)}$ represent in the context of this situation?

$\dfrac{\Delta \text{ pop}}{\Delta \text{ years}}$

  f. What does the expression $\dfrac{h(10)}{h(0)}$ represent in the context of this situation?

  g. What is the meaning of $5h(k)$ in the context of the situation?

*4. This table represents the price (in dollars) of a new Toyota Camry as a function of the number of years, n, after 2000.
  a. What is the input quantity? What is the output quantity?

# years since 2000, n	Price of car, $f(n)$
0	17,520
1	17,675
2	18,970
3	19,045
4	19,875
5	19,295
6	19,545
7	19,925

  b. Evaluate each expression and describe the meaning in the context of the problem.
    i. $f(7) =$

    ii. $2f(4) =$

    iii. $f(7) + 1250 =$

    iv. $f(6) - f(3) =$

    v. $\dfrac{f(6)}{f(5)} =$

*5. Suppose the following graph of the function f represents John's weight (in pounds) as a function of time t, measured in days since January 1, 2008.

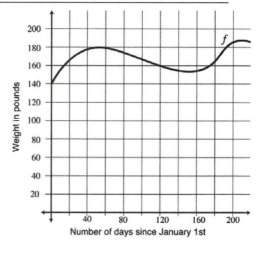

a. Identify the input and output quantities for the function f.

In: Time   Output = Weight

b. Evaluate $f(60)$. What does this value represent in the context of the problem?

60 days after Jan, 1 2008

c. Solve $f(x) = 160$ for x using the graph. Describe what each solution represents.

y= 18, 120, 178

d. What does $f(0) = 140$ represent in the context of this situation?

he weighd 140 lbs on Jan 1 2008

e. Evaluate $f(200) - f(160)$. What does this difference represent in the context of the problem?

f. Using function notation, construct an expression that represents a constant rate of change (the average rate of change) of John's weight from 160 days past January 1st to 200 days past January 1st? What could this constant rate of change mean in the context of this problem?

6. Evaluate each of the following:

a. $f(2x+3)$ when $f(r) = \dfrac{r}{4} + 2r^2 - 6r$.

b. $g(4) - g(2.1)$ when $g(p) = \dfrac{2\sqrt{p+14}}{8-p} + 5$.

c. $h(2) + k(3)$ when $h(x) = 4^x + 2x - 7$ and $k(y) = \dfrac{4}{9}\sqrt{y^4 - 3y} - 5.1$

*7. Given the existence of some functions t and b,
a. Use function notation to represent the change in the output value of function b over the input interval from -3 to 6.

b. Use function notation to represent the sum of the function t at an input of -4 and the function b at an input of -1.3.

c. Use function notation to represent the average rate of change of the function b from an input of 5 to an input of 9.

*1. Running is a popular form of exercise to burn calories and stay healthy. The amount of calories burned while running depends on many factors, but averages about 100 calories per mile. Suppose Nikki goes for a run, traveling at a constant speed of 720 feet per minute and burning 100 calories per mile she runs.

    a. What quantities are varying (changing) in this situation? What quantities are constant? Be sure to include units for each quantity.

    b. Is it possible to write a formula to define the number of calories burned in terms of the number of minutes that Nikki runs? If so, describe the process or processes, along with the input and output variables for each process, that you could use to determine the number of calories burned if Nikki runs 50 minutes.

    c. Describe how you think each of the following pairs of quantities are changing together.
        i. As the time (in minutes) spent running increases, how does the distance (in feet) Nikki has traveled change?

        ii. As the distance (in feet) Nikki has traveled increases, how does the number of calories she has burned change?

        iii. As the time spent running increases, what happens to the number of calories Nikki has burned?

    d. Complete the following table of values for this situation. (Recall there are 5280 feet in one mile.)

Time (in minutes) Nikki has been running	Distance (in feet) Nikki has traveled	Distance (in miles) Nikki has traveled
3		
14		
18.5		
22.2		
t		

    e. How many calories does Nikki burn if she runs 3 miles? 5.7 miles? m miles?

f.   Complete the following table of values relating the number of minutes Nikki spends running and the number of calories she burn.

Time (in minutes) Nikki has been running	Number of calories Nikki has burned
4	
11	
15.5	
19.2	

g.   How many calories will Nikki burn if she spends t minutes running?

*2. Suppose a square spontaneously appears and begins to grow continuously so that the length of each side of the square grows at a constant rate of 3 inches per second (assume that the side length of the square is 0 inches when the square has been growing for 0 seconds).
a.   Illustrate this situation with a diagram.

b.   Define a function g that determines the side length (in inches) of the square s in terms of the number of seconds t since the square started expanding from a side length of 0 inches. The expression $g(t)$ represents the values of what quantity?

c.   How does the side length of the square change as the amount of time the square has been growing increases from 1 second to 4 seconds? Represent this change in side length using function notation.

d.   What does the expression $3g(3)$ represent in this context? What about $g(2)^2$?

e.   Define a function h that determines the area of the square in terms of the side length of the square, $s = g(t)$. (Recall that the area of a square is the number of 1-by-1 square units that "fit" inside the square—the area of a square is determined by the formula $A = s^2$ where s is the square's side length.)

f.   Define a function f to determine the area of the square in terms of t, the number of seconds since the square started expanding.

g.   Using function notation, represent the area of the square (in inches^2) when it has been expanding for 2 seconds, 3 seconds, 4 seconds. Next, compute these values of the area of the square. Is the area of the square increasing at a constant rate of change? Explain.

*3. Michael and Cameron are both riding their bikes on separate routes to raise money for a *Clean Air Campaign*. They begin riding at the same time. Michael rides at a constant rate of 15 miles per hour and Cameron rides at a constant rate of 17 miles per hour.

a. Define a function *f* that determines the combined distance traveled (in miles) by both Michael and Cameron in terms of *t*, the number of hours since they started riding.

b. For every 7 miles they ride they raise $2 for the *Clean Air Campaign*. Define a function *g* to determine how much money they have raised in terms of the number of hours *t* they ride.

*4. Use the table of values provided to answer the following questions.

a. *g*(*f*(−1)) = 0

b. *f*(*f*(3)) = 0

c. *g*(*g*(0)) = −1

d. *f*(*g*(3)) = 4

e. If *f*(*g*(*x*)) = 3, then what must the value of *x* be? = x=2

x	*f(x)*	*g(x)*
−2	0	5
−1	3	3
0	4	2
1	−1	1
2	6	−1
3	−2	0

*5. The following graphs show two functions, *f* and *g*. Function *g* takes as its input a temperature in degrees Fahrenheit and outputs the expected attendance at a neighborhood carnival. Function *f* takes as its input a number of people attending the carnival and outputs the total expected revenue earned by the carnival.

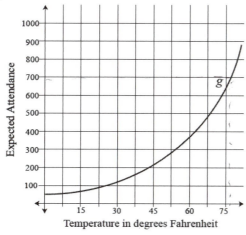
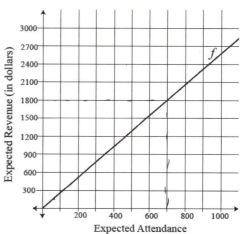

a. Does the expression *f*(*g*(70)) have a real-world meaning in this context? If so, estimate its value and explain what the value represents. If not, explain your reasoning.

$1250 are generaly earned when it is 70°F out and around 490 people are present

b. Does the expression *g*(*f*(70)) have a real-world meaning in this context? If so, estimate its value and explain what the value represents. If not, explain your reasoning.

NO, because the functions of money and temp cannot be reled difectly related

c. Let's define a new function k that is the composition of f and g, that is, $k(x) = f(g(x))$. Explain what the equation $1800 = k(x)$ represents, then explain how you can find the value of x that satisfies the equation.

$g(x) = 77$ $K(x)$ represents revenue in dollars

6. A pebble is thrown into a lake and the radius of the circular ripple increases at a constant rate of 0.7 meters/second. Your goal is to determine the area (in meters^2) inside the ripple in terms of the number of seconds elapsed since the pebble hit the water. Before determining a function to represent this relationship, discuss the following questions in your group.

   a. Draw a picture of the situation and label the quantities. Imagine how the quantities are changing together. Discuss in your groups what processes need to be carried out to determine the area inside the ripple when the number of seconds since the pebble hit the water is known.

   b. What quantities are varying (changing) in the situation and how are they changing together?
      i. As the time since the pebble hit the water increases how does the radius of the ripple change?

      ii. As the radius of the circular ripple increases how does the area of the ripple change?

      iii. As the time since the pebble hit the water increases how does the area of the circular ripple change?

   c. Define variables to represent the values that the relevant varying quantities can take on. The goal described originally was to relate what quantities?

   d. Define the following functions:
      i. f that defines the radius (in meters) of the circular ripple r as a function of the time elapsed (in seconds) t since the pebble hit the water.

      ii. g that defines the area of the circular ripple A (in meters^2) as a function of the radius (in meters) of the circular ripple r.

© 2015 Carlson, Oehrtman, and Moore

e. Graph *f* and *g* and use these graphs to determine the area 4 seconds after the pebble hits the water. Explain your process.

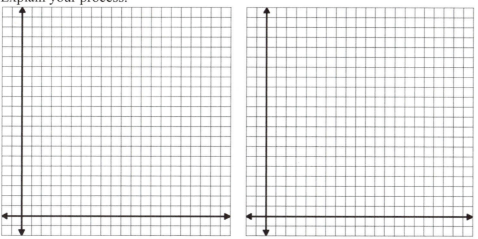

f. Use the functions defined in part (d) to define a function *h* that defines the area of the circular ripple *A* (measured in meters^2) as a function of time elapsed *t* (measured in seconds).

g. Compute the value of *h*(4). How does this answer relate to the answer you obtained in part (e)?

7. Use the graph provided to approximate answers the following questions.

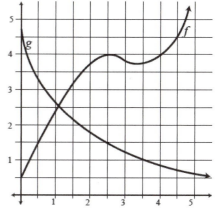

a. *f*(*g*(2.5))   4

b. *g*(*f*(4))

c. *g*(*g*(3.5))

d. Determine the value(s) of *x* such that *f*(*g*(*x*)) = 2.5

8. Use the following functions to answer the questions below: $f(x)=\sqrt{x+3}$, $g(x)=2x+9$, $h(x)=\frac{x}{4}$.

a. Evaluate *f*(*g*(2)).   4

b. Evaluate *h*(*f*(61)).

c. Function *m* is defined as *m*(*x*) = *g*(*h*(*x*)). Write the rule for *m*.

d. Function *p* is defined as *p*(*x*) = *g*(*f*(*x*)). Write the rule for *p*.

1. Tim works 30 hours a week between two part-time jobs: waiter ($9.25 per hour) and math tutor ($8.50 per hour). Since he has only 30 hours each week that he can work, the more hours he spends at one job the fewer hours remain for the second job.

   a. Assume that Tim worked a total of 30 hours this week and he worked 9 hours as a waiter. How much money in total did Tim make this week? Explain how you determined your answer.

   b. Our goal is to define a function that determines the total amount of money Tim makes when he works x hours as a waiter.

      i. Define a function h that determines the amount of money Tim makes for working for y hours as a math tutor.

      ii. Define a function, k, to determine the value of y, the number of hours Tim works as a math tutor in terms of the number of hours x Tim works as a waiter.

      iii. Define a function g that determines the amount of money Tim makes for working as a math tutor when he works for x hours as a waiter.

      iv. Define a function f that determines the amount of money Tim makes for working for x hours as a waiter.

      v. Define a function t that determines the total amount of money Tim makes in a given week working both jobs in terms of the number of hours, x, he works as a waiter.

   c. i. Graph functions g, f, and t on the same axes.

      ii. Use your graph to evaluate $g(12) + f(12)$. What do you notice? What does the expression $g(12) + f(12)$ represent?

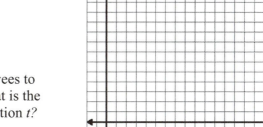

   d. The restaurant Tim works at requires its employees to work between 15 and 25 hours each week. What is the domain of function t? What is the range of function t?

   e. Suppose Tim must work at least 12 hours each week as a math tutor. How does this change the domain and range you determined in part (c)?

*2. a.   What is the radius of a circle with a circumference of 15 feet?

b.   What is the area of a circle with a radius of 2.387 feet?

c.   Describe a two-step process to determine the area of a circle when the circumference of the circle is known.

d.   Use the two-step process described in part (c) to determine the area of a circle when the circumference of the circle is 29.63 feet.

e.   Define a function, *f*, to determine the radius of the circle given the circumference of the circle.

f.   Define a function *h* that determines the area of a circle in terms of the circumference of the circle.

g.   Use function *h* defined in part (f) to determine the area of a circle when the circumference of the circle is 29.63 feet. How does your answer compare to your answer in part (c)?

*3.   The following graphs show two functions, *f* and *g*. Function *g* takes as its input the side length of a square and outputs the area of the square. The function *f* takes as its input the side length of a square and outputs the perimeter of the square.

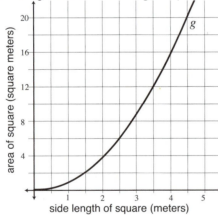
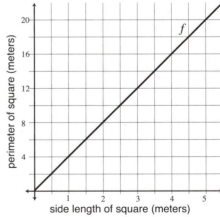

a.   Does the expression *f*(*g*(4)) have a real-world meaning in this context? If so, estimate its value and explain what the value represents. If not, explain your reasoning.

No, you would be going ~~area~~ to ~~inputs~~ units of perimiter ~~to~~ from the side lengths of squres.

b. Does the expression $g(f(4))$ have a real-world meaning in this context? If so, estimate its value and explain what the value represents. If not, explain your reasoning.

No, you would be going ~~from~~ to units of the ~~side~~ area of a square to the side length

c. Is it possible, using the graphs provided above, to determine the area of a square given the when the perimeter of the square is 4 meters? If so, explain how. If not, explain why.

Yes, both graphs use a constant unit of side length of squares so when the the perimeter = 4m the side length is 1, when the side length is 1, the area is 1 sq m

d. Our goal is to determine the area of the square given the perimeter of the square.
  i. What information do you need to know in order to determine the area of the square?
  The relationship between Perimeter and side length and area and side length

  ii. Using the graphs, determine the area of the square when the perimeter of the square is 8 feet.

  4 sq m²

  iii. If we know that $A = s^2$ and $p = 4s$, define a formula that gives the area of the square in terms of its perimeter.
  $\sqrt{A} = s = \dfrac{P}{4}$

*4. A farmer has 250 feet of fencing to create a rectangular pen.
  a. Draw a picture of the situation and label the relevant quantities.

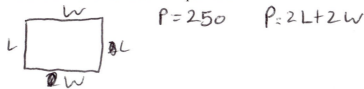

  $P = 250$ $P = 2L + 2W$

  b. If the width w of the pen is 40 feet, then what is the length of the pen l? (Hint: Recall that the perimeter p of a rectangle is $p = 2l + 2w$) $250 = 80 + 2L$ $170 = 2L$
                                                                                                              $L = 85$

  c. If the width of the pen is w feet, what is the length of the pen, l? (Hint: note that the perimeter for this situation is a fixed value.) $\dfrac{P}{2} = L + w$ $L = \dfrac{P}{2} - w$

  d. Define a function f that expresses that area of the rectangular pen (measured in square feet) as a function of w, the width of the rectangular pen (measured in feet).

  $f(w) = \left(\dfrac{P}{2} - w\right) w$

  $f(w) = \dfrac{250 - 2w}{2} \circ w$ $f(w)$
                                          $u(z)$

5. A bakery is known for their chocolate cakes. Let *m* represent the number of cakes made and let *s* represent the number of cakes sold. Their weekly profit in dollars, *P*, is given by the formula $P = 22s - 9m$.

   a. Assume that the bakery has a weekly profit of $515 and made 65 cakes that week. How many cakes were sold that week?

$$515 = 22s - 9(65)$$
$$+ 585$$

$$1100$$
$$+ \cancel{two} = 22s$$
$$s = \cancel{100}\,50$$

   b. Suppose the weekly profit is $515 and *m* cakes were made at the bakery that week. Write a formula that determines how many were sold that week.

$$515 = 22s \cdot$$

$$P = 22s - 9m$$
$$\frac{P + 9m}{22} = s$$

   c. Chanel is the only baker in the shop and she gets paid based off the number of cakes she makes and the number of cakes she sells. Assume that she makes $10 for every cake she makes and $5 for every cake she sells. Write a formula that gives her commission, *C*, in terms of the number of cakes she makes, *m*, and the number of cakes she sells, *s*.

$$10m + 5s = C$$

   d. Write a function, *t*, that gives Chanel's commission, *C*, in terms of the number of cakes she makes in a week. (Assume that the weekly profit is $515 and recall that Chanel is the only baker at the bakery).

$$f(t) =$$

$$C(t) = 10t + 5\left(\frac{\cancel{515}\,515 + 9t}{22}\right)$$

The idea of function inverse is not new to you—it is nothing more than an ***undoing*** *(*or reversal) of the function rule or process.

Let's consider a simple example that we will build upon in problem 1:

Suppose water is being added to an empty swimming pool at a rate of 7 gallons per minute. If we have a function *f* that determines the number of gallons of water in the pool in terms of the number of minutes *t* since the pool started filling, we can define this function as $f(t) = 7t$. This function *f* determines the number of gallons of water in the pool (dependent quantity) "in terms of" the time *t* in minutes since water started to be added to the empty pool (independent quantity).

When we evaluate the function *f* for a value of $t = 6$, we write $f(6) = 7(6) = 42$, which means that when 6 minutes have passed since the pool started filling, there are 42 gallons of water in the pool.

1. a. Read the example above. Now, let's think about reversing the rule of *f* and consider:
    i. the amount of time *t* in minutes that water has run in, when 84 gallons are in the pool

    ii. the amount of time *t* in minutes that water has run in, when 168 gallons are in the pool

    iii. the amount of time *t* in minutes that water has run in, when 15 gallons are in the pool

   b. Solve $35 = 7t$ and say what the answer means.

   c. We see that solving an equation for the independent variable *t* when a specific value of the dependent variable $f(t)$ or *v* is known, involves a reversal (or undoing) of the process defined by *f*.
    i. Evaluate $f(12)$ and describe what the answer represents.

    ii. Solve $84 = 7t$ and describe what the answer represents.

    iii. What do you observe? How are ideas of evaluating a function and solving an equation for the independent value of that function related?

   d. Consider a function *g* that determines the time (in minutes) the pool has been filling in terms of the volume *v* (in gallons) in the pool.
    i. What is the independent quantity?

    ii. What is the dependent quantity?

   e. Define the function *g* that determines the time (in minutes) the pool has been filling in terms of the volume *v* (in gallons) in the pool.

   f. Evaluate *f* when $t = 9$ minutes, then evaluate *g* when $v = 63$ gallons

We say the function *g* is the inverse function of *f* because it undoes the rule of *f*. It is standard to reference a function that is the inverse function of a specific function *f* with the notation f^{-1}. We express that f^{-1} (read "*f* inverse") and *g* are the same functions by writing, $g = f^{-1}$. More specifically this conveys that the function rules for these two functions *g* and f^{-1} produce exactly the same value for their dependent quantity when evaluating the same values of the independent quantity (the rules of the two functions are equivalent).

2. Explain what it means for two functions *h* and *m* to be equal.

3. The function *h* determines the circumference *C* of a circle in terms of its radius *r*. Because the circumference of any circle is 2π (or about 6.28) times as large as its radius *r*, the formula $C = 2\pi r$ expresses the circumference in terms of the radius and define the function *h* by writing $h(r) = 2\pi r$.
   a. What does *h*(4) represent?

   b. Define h^{-1} and describe its independent quantity and its dependent quantity.

   c. What conjecture do you make about $h^{-1}(h(r))$? (Assume *r* and *h*(*r*) are positive real numbers.) Support your conjecture by referencing the quantities in the situation.

A new notation has been introduced by mathematicians to name a function that undoes the rule (or reverses the process) of a function *f*. This notation is f^{-1}. The independent (or input) quantity of *f* is the dependent (or output) quantity of f^{-1}, and the independent quantity of f^{-1} is the output quantity of *f*.

Even though the examples in Exercise 1 – 3 are very simple, the idea of a function *f* and its inverse function f^{-1} as undoing each other applies the same for all functions, no matter how complex. We will explore later in this investigation situations when a function *f* does not have an inverse function f^{-1}. However, if both *f* and f^{-1} are functions that are defined for all real numbers (guaranteeing that the domain of f^{-1} is equal to the range of *f* and the domain of *f* is equal to the range of f^{-1}), then

$$f\left(f^{-1}(y)\right) = y \text{ and } f^{-1}(f(x)) = x$$

*4. When traveling outside of the United States it is often useful to be able to convert between temperatures measured using the Fahrenheit scale and temperatures measured using the Celsius scale. The standard formula for determining temperature in degrees Fahrenheit when given the temperature in degrees Celsius is, $F = \frac{9}{5}C + 32$. We can write this formula using function notation by letting $F = g(C)$ and writing $g(C) = \frac{9}{5}C + 32$. The function *g* defines a process for converting degrees Celsius to degrees Fahrenheit.
   a. State the meaning of *g*(100) and evaluate *g*(100).

   b. Solve the equation *g*(*C*)=212 and explain how you arrived at your answer.

   c. Define a function *h* that converts Fahrenheit temperature to its Celsius temperature. (Hint: Generalize the steps you described in part (b) that reversed the process of *g*. You can also solve $F = \frac{9}{5}C + 32$ for *C*.)

---

d. Determine the values of $g(100)$ and $h(212)$.

e. Determine the values of $g(h(212))$ and $h(g(100))$ without performing any calculations. What do you notice about the relationship between g and h?

f. Represent the value of $g(h(n))$. Represent the value of $h(g(k))$. What do you notice about the relationship between g and h?

> Two functions are said to be **inverses** of each other if composing the two functions gives you the value you start with. So, two functions u and v are inverses of each other when $u(v(n))=n$ and $v(u(m))=m$ for any values of n and m that make sense in the context of the functions u and v. So, the functions for converting from degrees Celsius to degrees Fahrenheit and from degrees Fahrenheit to degrees Celsius are inverses of each other. Another way to think of the functions that are inverses of each other is that each function undoes what the other function does to its input.
>
> A function that is the inverse of a function f is represented as f^{-1}. Note f^{-1} DOES NOT MEAN $\frac{1}{f}$. Be careful to not confuse the inverse of the function f (a function that undoes the process of f) with the idea of inverting a number. Recall that the inverse of a number such as 2 is written $2^{-1} = \frac{1}{2}$ and the inverse of a variable x is written $x^{-1} = \frac{1}{x}$.

*5. Use the functions h and g defined in Exercise #4.
   a. Is it correct to say that the function h undoes the process defined by the function g? Explain.

   b. How would you represent the fact that h is the inverse function of g?

*6. Use the given table to evaluate the following.
   a. $g^{-1}(3)$ $\quad -1$

   b. $g^{-1}(0)$ $\quad 3$

   c. $f(g(0))$ $\quad \downarrow$

   d. $f^{-1}(-1)$ $\quad 1$

   e. $g^{-1}\left(f^{-1}(6)\right)$

   f. $f^{-1}\left(g(2)\right)$ $\quad 1$

x	$f(x)$	$g(x)$
−2	0	5
−1	3	3
0	4	2
1	−1	1
2	6	−1
3	−2	0

*7. Given the function f defined by $f(x) = x^2$, we have $f(2) = 4$ and $f(-2) = 4$, and know each output value of f greater than 0 is paired with two input values (see graph). Evaluate f^{-1} for the input value 4. That is, evaluate $f^{-1}(4)$. What do you observe?

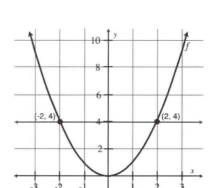

Recall that the definition of a function requires that each input from a function's domain is paired with exactly one output value in that function's range. Since inputting 4 to f^{-1} returns output values 2 and -2, we conclude that f^{-1} is **not** a function.

8. Given $f(r) = 4\pi r^2$ is the function whose input is the radius of a sphere in inches and whose output is the surface area of the sphere in square inches.

   a. Evaluate $f^{-1}(115)$. Explain the meaning of your answer.

   b. Define the inverse function f^{-1} that undoes the process of f. Describe the process, input quantity, and output quantity of f^{-1}.

   c. Explain what it means to evaluate $f^{-1}(26.7)$.

   d. Explain what it means to solve $f(x) = 26.7$ for x.

   e. Compare your answers to part (c) and (d). What do you notice? Does this make sense? Explain.

9. The function f gives the volume of a sphere as a function of its radius length, $f(r) = \frac{4}{3}\pi r^3$
   a. What is the domain of f?

   b. What is the range of f?

   *c. Determine a formula for the inverse of the function f. What is the input quantity of f^{-1}? What is the output quantity of f^{-1}?

   *d. What is the domain of f^{-1}? What is the range of f^{-1}?

10. Determine a formula that undoes the process of each of the given formulas.

    a. $A = f(s) = s^2$

    b. $d = f(t) = 56t$

    c. $y = f(x) = x - 8$

    *d. $A = f(r) = \pi r^2$

    *e. $y = f(x) = \frac{2x-5}{6}$

This investigation revisits an idea that has received a good deal of focus so far in this course—the idea of determining a constant rate of change (of the dependent quantity with respect to the independent quantity) over a specified interval of the independent quantity.

1. Given a function f defined by the formula $f(t) = -4t^2 + 12t + 2$ determines the height (in meters) of a ball above the ground t seconds after you tossed it straight up into the air.
   a. Use function notation to express the increase in the height of the ball above the ground, as the amount of time t since you threw it increased from 0.5 seconds to 1.5 seconds.

   b. Construct a graph of f, determine the slope of the line connecting the points $(0.5, f(0.5))$ and $(1.5, f(1.5))$, and then construct the line.

   c. Describe what this constant rate of change could represent in the context of this situation.

   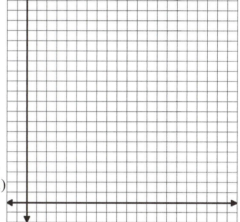

   d. Use function notation to write an expression to determine the constant rate of change between any two points $(t_1, f(t_1))$ and $(t_2, f(t_2))$ on the graph of f, given that $0 \le t_1 \le 3.15$ and $0 \le t_2 \le 3.15$. (The expression is sometimes referred to as the difference quotient.)

2. Recall that the volume of a box V (measured in cubic inches) as a function of the length of the side of the square cutout x (measured in inches) is given by the function:
   $f(x) = x(11 - 2x)(8.5 - 2x)$.
   a. Describe the meaning of each of the following expressions in the context of the situation:
      i. $f(x+3)$ ii. $f(x+3) - f(x)$ iii. $\frac{f(x+3) - f(x)}{(x+3) - x}$

   b. Evaluate $\frac{f(x+3) - f(x)}{(x+3) - x}$ when $x = 0.5$. Describe the meaning of this value in the context of the situation.

3. The area A (measured in square feet) of a circular oil slick as a function of the amount of time (measured in minutes) since the oil leak started is given by the function $g(t) = \pi(7.84t)^2$.
   a. Describe the meaning of $\frac{g(t+5) - g(t)}{5}$ (simplified from $\frac{g(t+5) - g(t)}{(t+5) - t}$) in the context of this situation.

   b. Evaluate $\frac{g(t+5) - g(t)}{5}$ when $t = 1.5$. Describe the meaning of this value.

In general, the expression $\frac{f(x+h)-f(x)}{h}$ (simplified from $\frac{f(x+h)-f(x)}{(x+h)-x}$) where h represents the change in x is called the ***difference quotient***. The difference quotient is the average rate of change for a function between two input-output pairs (see point A to point B, as expressed in the graph). Using function notation, we say that you can find the average rate of change between any two points $(x, f(x))$ and $(x+h, f(x+h))$ by computing: $\frac{\text{change in outputs}}{\text{change in inputs}} = \frac{f(x+h)-f(x)}{(x+h)-x} = \frac{f(x+h)-f(x)}{h}$. Thus, when finding the difference quotient for two points on a function, you are determining the average rate of change between those two input-output pairs.

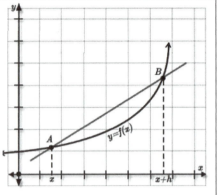

3. a. Find and interpret the difference quotient (average rate of change) for the population of a small town P (measured in a number of people) as a function of the number of years t since 1990 as defined by the function $f(t) = 1500 + 60t$.

   b. Find and interpret the difference quotient (average rate of change) for the distance a car is from an intersection d (measured in feet) as a function of time t (measured in seconds) as defined by the function $m(t) = 2t^2 + 4t + 1$.

4. Use the graph of $y = f(x)$ and an input value of $x = k$ to represent the following quantities on the graph.
   a. $f(k)$ b. $k + 2.5$

   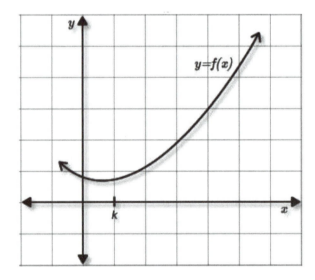

   c. $f(k + 2.5)$ d. $f(k+2.5) - f(k)$

   e. Represent the quantity $\frac{f(k+2.5)-f(k)}{2.5}$ on the graph. Explain how your representation conveys the value of this quantity.

I. THE BOX PROBLEM AND MODELING RELATIONSHIPS (Text: S1, S2)

1. A box designer has been charged with the task of determining the volume of various boxes that can be constructed by cutting four equal-sized square corners from a 14-inch by 17-inch sheet of cardboard and folding up the sides.
    a. What quantities vary in this situation? What quantities remain constant?
    b. Create an illustration to represent the situation and label the relevant quantities.
    c. What are the dimensions of the box (length, width, and height) if the length of the side of the square cutout is 0.5 inches? 1 inch? 2 inches?
    d. Define a formula to relate the height of the box and the length of the side of the square cutout. Be sure to define your variables.
    e. Let x represent the length of the side of the square cutout; let w represent the width of the base of the box; let l represent the length of the base of the box; and let V represent the volume of the box. Complete the following table assuming each quantity is measured in a number of inches (volume measured in cubic inches).

x	w	l	V
0	14	17	0
4			
9			
		2	
	0		

   f. Define a formula to relate the length of the base of the box and the length of the side of the square cutout. Be sure to define your variables.
   g. Use the formula created in part (f) to determine the average rate of change of the length of the base of the box as the length of the side of the square cutout increases from 1 inch to 3 inches.
   h. Define a formula to relate the volume of the box to the length of the side of the square cutout.
   i. Use the formula created in part (h) to determine the change in the volume of the box as the length of the side of the square cutout increases from 0.5 inches to 1 inch and from 1 inch to 1.5 inches. Why are these values not the same?
   j. Use your graphing calculator to approximate the maximum volume of the box rounded to the nearest tenth of a cubic inch. Explain how you know that the value you obtained is the maximum value of the volume of the box rounded to the nearest tenth.

2. A box designer has been charged with the task of determining the surface area of various open boxes (no lids) that can be constructed by cutting four equal-sized surface corners from an 8-inch by 11.5-inch sheet of cardboard and folding up the sides. (Note: The surface area is the total area of the box's sides and bottom.)
    a. What quantities vary in this situation? What quantities remain constant?
    b. Create an illustration to represent the situation and label the relevant quantities.
    c. Complete the following table

Cutout length (in)	Height of box (in)	Length of box (in)	Width of box (in)
0.5			
1			
2			

(problem continues on next page)

    d. Define a formula for the following relationships:
        i.   the length of the base of the box to the length of the side of the square cutout
        ii. the width of the box to the length of the side of the square cutout
    e. Define a formula that relates the total surface area, s, (measured in square inches) of the open box to the size of the square cutout x (measured in inches).
    f. There are two ways to think of determining a box's surface area: by (1) adding up the area of each piece of the box; and by (2) subtracting the area of the four cutouts from the area of the initial sheet of cardboard.
        i.   Which method did you use in defining your formula in part (e).
        ii. Define a formula to compute the box's surface area by the other method. Compare the two formulas. Are they equivalent (that is, do they always produce the same result)? Show this algebraically.

3. An open box is constructed by cutting four equal-sized square corners from a 10-inch by 13-inch sheet of cardboard and folding up the sides. The graph below represents the volume of a box (measured in cubic inches) in relation to the length of the side of the square cutout (measured in inches).

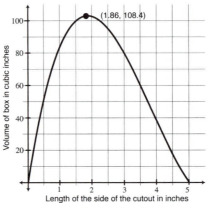

    a. Define a function g to relate the volume of the box to the length of the side of the square cutout.
    b. Determine the volume of the box when the length of the side of the square cutout is:
        i. 0 inches   ii. 1 inch   iii. 2.5 inches   iv. 5 inches
    c. Use the graph above to approximate the value(s) of the length of the side of the square cutout when the volume is:
        i. 0 cubic inches   ii. 100 cubic inches   iii. 108.4 cubic inches
    d. As the length of the side of the square cutout increases from 0 to 1 inch, by how much does the volume change? Illustrate this change on the graph.
    e. Use the graph to determine for what values of the length of the side of the square cutout the volume is increasing.
    f. Use the graph to determine for what values of the length of the side of the square cutout the volume is decreasing.

For Problems 4 – 6: Windows, Inc. manufactures specialty windows. One of their styles is in the shape of a semicircle, as shown below. The cost of manufacturing the window consists of the cost of the glass pane together with the cost of the framing material.

l

4. When a customer orders this window, they specify the length of the base of the window.
   a. Determine the total length of frame needed for a window with a base of 4 feet.
   b. Determine a function named h that relates the total length of the frame to the length of the base of the window.
   c. If the cost of the framing material is $12 per linear foot, what is the cost of a window frame with a base-length of 4 feet?
   d. Suppose that your budget limits you to spending $500 on your window frame. What is the longest base length that you can afford?

5. Suppose that the glass pane costs $23 per square feet.
   a. What is the area of a glass pane when the base of the window is 6 feet?
   b. Write a function named k that relates the area of the glass pane to the length of the base of the window.
   c. What is the cost of the glass pane when the base of the window is 6 feet?
   d. Suppose that your budget limits you to spending $850 on your glass pane. What is the longest base length of the window that you can afford?

6. The total cost of the window includes both the cost of the glass pane at $23 per square foot and the cost of the window frame at $12 per linear foot. What is the total cost of a window with a base of 3 feet? (Hint: Consider the functions you defined in problems 4 and 5.)

7. The expression $224 - 4x^2$ calculates the surface area of a box (measured in square inches) when the box is made with a square cutout of side length x (measured in inches) from a 14" by 16" sheet of paper.
   a. Define a function, named g, that has the length of the square cutout x as input (measured in inches) and the surface area A of the corresponding box as output (measured in square inches). Label the expressions in your function's definition.
   b. Using function notation, represent the surface area of the box when the length of the side of the square cutout is 0.2 inches, 2.7 inches, and 4.1 inches. (That is, *do not* calculate the surface area. Instead *represent* the surface area determined by each length of the side of the square cutout.)
   c. What does the expression $g(1.3)$ represent?

8. The expression $\frac{4}{3}\pi r^3$ calculates the volume of a sphere (measured in cubic inches) when the radius of the sphere is r inches.
   a. Define a function, named h, that has the radius of the sphere, r, as input (measured in inches) and the volume of the sphere, V, as output (measured in cubic inches). Label the expressions in your function's definition.
   b. Using function notation represent the volume of the sphere when the radius of the sphere is 2.4 inches, 3.1 inches, and 5.2 inches. (That is, *do not* calculate the volume. Instead *represent* the volume determined by each radius length.)
   c. What does the expression $h(19.6)$ represent?

Instructions for Problems 9 – 14: Evaluate the expression.

9. Find the value of $\dfrac{(9+x)-x^2}{2x+7}$ when $x = 3$.

10. Find the value of $\dfrac{3x^2 + x - 2(3x+5)}{2x+9}$ when $x = 8$.

11. Find the value of $\dfrac{\left(\dfrac{30}{y} + \dfrac{49}{y+2} + \dfrac{100}{y^2}\right)}{\left(\dfrac{20}{2y}\right)}$ when $y = 5$.

12. Find the value of $\dfrac{(x+4)^3 - 4y + 6xy}{-5x+6y}$ when $x = 2$ and $y = 0.5$

13. Find the value of $y\sqrt{x+4} - 3x - 7y + \dfrac{4x}{8}$ when $x = 12$ and $y = -4$

14. Find the value of $\left(2(x-7)\right)^2 - \dfrac{y}{6}$ when $x = -2$ and $y = 54$.

II. FUNCTION RELATIONS AND DOMAIN OF FUNCTIONS (Text: S1)

Instructions for Problems 15 – 20:
   a. Determine the input,
   b. Determine the output,
   c. Determine if the relation is a function.

15. The cost of a pound of produce at a specific store with respect to the barcode number of that type of produce

16. The barcode number of a type of produce with respect to the cost of a pound of produce at a specific store.

17. A student's ID number with respect to the year of that student's birth.

18. The year of a student's birth in terms of the student's ID number

19. In a given apartment complex, the apartment number in terms of the number of people living in that apartment.

20. In a given apartment complex, the number of people living in that apartment in terms of the apartment number.

For Problems 21 – 22 use the following table.

Name of Person	John	Mary	Sally	Michael	Dawn
Height of Person (inches)	71	63	64.5	71	68

21. Is the relation the height of a person in terms of the name of the person a function? Explain your reasoning.

22. Is the relation the name of the person in terms of the height of a person a function? Explain your reasoning.

23. For each of the relations below, determine whether y is a function of x. Use the definition of a function to justify your answer.

a.

x	y
1	3
2	2
1.7	4.5
2.1	9
2	1.1
5	7

b.

x	y
1	3
-2	-1
3.7	0.4
1.5	-5
3.7	0.4
-4	-1

c.

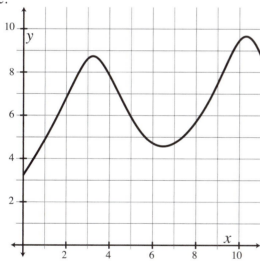

d.

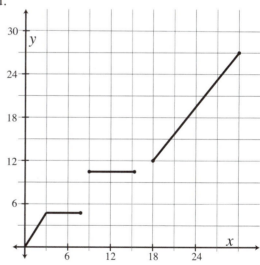

e. $y = x(8.5 - 2x)(11 - 2x)$

f. $y = 2^x$

g. A manufacturing plant produces bags of plaster weighing between 0 and 90 pounds. Bags weighing up to 50 pounds are marked with a "1" (indicating that the bags can safely be carried by one person), and bags weighing more than 50 pounds are marked with a "2" (indicating that the bags are heavy and should be carried by two people). Is the number marked on the bag y a function of the weight of the bag x?

24. For each of the relations in Question 23, determine whether x is a function of y. Justify your answers.

25. Create two tables that represent relations between two variables that are functions. Explain why one variable is a function of the other variable.

26. Create two tables that represent relations between two variables that are NOT functions. Explain why the relationships are not functions.

27. Determine the domain of the following functions.
    a. $f(x) = 3x^2 + 5x - 7$

    b. $h(x) = \dfrac{x}{5x - 10}$

    c. $k(x) = \sqrt{x + 2}$

    d. $n(x) = \dfrac{\sqrt{x - 7}}{x^2 - 4}$

28. Determine the domain of the following functions.
    a. $g(x) = 5x^6 + 2x^8 - 3x^3 + 4x - 1$

    b. $p(x) = \dfrac{2x + 3}{x - 4}$

    c. $s(x) = \sqrt{3x - 7}$

    d. $m(x) = \dfrac{x}{\sqrt{3x - 4}}$

III. Using and Interpreting Function Notation (Text: S3, S4)

29. The expression $224 - 4x^2$ calculates the surface area of a box (measured in square inches) when the box is made with a square cutout of side length x (measured in inches) from a 14" by 16" sheet of paper. Let g be a function that takes as its input the length of the side of the square cutout (measured in inches) and outputs the surface area of the corresponding box (measured in square inches).
    a. What does the expression $g(1.5) - g(0.5)$ represent in the context of the problem?
    b. What does the expression $g(6.25) - g(2.1)$ represent in the context of the problem? Calculate the value of $g(6.25) - g(2.1)$.
    c. What does the expression $4g(3.5)$ represent in the context of the problem?

30. Functions are commonly used by computer programmers to access databases. All databases that involve time have "reference times". Suppose a database uses a function named g to access the population of Loveland, CO. This database uses 12:00am, January 1, 1990 as its reference time. So, $P = g(u)$ represents Loveland's population u years from 12:00am, January 1, 1990.
    a. Represent Loveland's population on 1/1/1982.
    b. Represent the change in Loveland's population from 1/1/1994 to 1/1/1997
    c. Represent 6 times Loveland's population on 1/1/1972
    d. What does $g(8) - g(-13)$ represent?
    e. What does $g(h + 3)$, where h is some number of years, represent?
    f. What does $g(h + 3) - g(3)$, where h is some number of years, represent?
    g. What does $\frac{g(h+3)-g(3)}{h}$, where h is some number of years, represent?

31. Suppose that the function *h* relates a dairy farmer's cost (measured in dollars) as a function of the number of gallons of milk produced. Explain what each of the following represents.
    a. $h(18)$
    b. $h(b) = 25$
    c. $h(15) - 7h(9)$
    d. $2h(5) - 1 = 7.50$
    e. $h(4.3) + h(6.4)$
    f. $5h(4.5) + 2h(3) = 24.23$

32. Use function notation to express each of the following.
    a. When the input to the function *q* is 10, the output is 12.
    b. The output of a function *h* is *c* when the input value is 34.
    c. The area of a square is the square of the length of one side.
        i. Define a function *f* that expresses the area of a square as a function of its side length *s*.
        ii. The area of a square is 25m² when the side length is 5m.
        iii. Write an expression using the function *f* that represents how much larger the area of a square with side length of 6 inches is than the area of a square with side length of 2 inches.
    d. The circumference of a circle is π times as large as the circle's diameter *d*.
        i. Define a function *g* to express the circumference of the circle in terms of the circle's diameter.
        ii. Three times the circumference of a circle that has a radius of length of 5 units is approximately 94.2.

33. The following table gives values of the circumference *C* of a circle (in centimeters) as a function of its radius *r* (in centimeters).

Radius, r, (cm)	Circumference $C = f(r)$, (cm)
1	6.283
1.5	9.425
2	12.566
2.5	15.708
3	18.850

   a. What does the expression $f(1.5)$ represent? What is the numerical value of $f(1.5)$?
   b. What does the expression $f(2) - f(1.5)$ represent? What is the numerical value of $f(2) - f(1.5)$.
   c. What does the expression $\frac{f(3)-f(1.5)}{1.5}$ represent? What is the numerical value of $\frac{f(3)-f(1.5)}{1.5}$ and explain what this value represents in the context of this situation.
   d. When the radius of a circle changes from 1.5 to 2.5 cm, by how much does the circle's circumference change?
   e. Express the solution to part (d) using function notation.

34. A local entrepreneur has decided to make and sell computers in his own store. The cost to build each computer is $550 and the initial cost of starting the store is $10,000. The entrepreneur has also decided to sell each computer at a price of $1199.99 (this is the store's revenue per computer). Because he makes a computer only after someone has ordered it, he sells every computer he builds.
   a. Define a function *f* that expresses the cost of the computer store as a function of the number of computers built by the entrepreneur. Be sure to use function notation, and define all relevant variables.
   b. Identify the input and output quantities of the cost function. Explain *why* this relationship is a function.

(problem continues on next page)

c. Define a function *g* that expresses the revenue (the amount of money received before any bills are paid) as a function of the number of computers sold. Be sure to use function notation and define all variables and constants used.

d. Identify the input and output quantities of the revenue function. Explain *why* this relationship is a function.

e. Use function notation to represent how much the store's revenue changes as the number of computers sold increases from 11 to 14.

f. Assume that there are no other costs and that the entrepreneur sells all the computers that he produces, define a function named *h* that expresses the store's profit (revenue minus cost) as a function of the number of computers built. Be sure to use function notation, define all relevant variables, and identify the input and output quantities of the function.

g. How many computers must the store sell in order to break even (neither lose money nor make a profit.

35. Given the existence of some function *b* and *k*:
   a. Use function notation to represent the change in the output of the function *k* from an input of 6 to an input of 9.
   b. Use function notation to represent the sum of the output of the function *k* at an input of –9 and the output of the function *b* at an input of 6.12
   c. Use function notation to represent the average rate of change of the function *k* from an input of 4.1 to 12.5

36. Use the following information to answer the questions below.
   $C(t)$ represents the number of cats owned by people living in the United States *t* years since 2000. $D(t)$ represents the number of dogs owned by people living in the United States *t* years since 2000.
   a. Represent the total number of cats and dogs $P(t)$ owned by people living in the United States *t* years since 2000.
   b. Represent, using function notation, how many more cats than dogs were owned by people living in the United States *t* years since 2000.

37. Evaluate each of the following:
   a. $f(13)$ when $f(x) = 4.5 - 6x$
   b. $f(5)$ when $f(x) = \dfrac{2x^2 + (6 - 4x)}{9 - 3x}$
   c. $g(2.6)$ when $g(x) = \dfrac{x^3 - 2.7x - 5(-2x + 4)}{-5.9 + 3.2x}$
   d. $m(-6.1)$ when $m(x) = \dfrac{(x + 4)^2 - (4.1x)^2 + 3x}{1.1 + 7x}$

38. Evaluate each of the following:
   a. $f(x + 2)$ when $f(x) = 4x^2 - 2x + 10$
   b. $h(2x)$ when $h(y) = \dfrac{y^3 - 2y^2 + 4}{2y}$

IV. Function Composition: Stringing Processes Together (Text: S5)

39. Reggie, a college student, works part-time during the school year. He has to budget his expenditures so that he has enough money at the end of the month to cover his rent, car payment, and insurance. Currently, Reggie can only afford $15 per week for gas.
    a. As the price of gas fluctuates, the amount of gas Reggie can purchase each week varies. Complete the following table of values showing the number of gallons of gas Reggie can purchase with $15 at the given fuel prices.

Price of fuel (in dollars per gallon)	Number of gallons of fuel Reggie can purchase for $15
3.199	
3.499	
3.599	
3.799	
p	

    b. Reggie's car gets an average of 28 miles per gallon. How many miles can he drive in a week if he purchases 3 gallons of fuel? 4.18 gallons? g gallons?
    c. Explain how you can determine the number of miles Reggie can drive in a week if gas costs $3.899 per gallon and he has $15 to spend on gas.
    d. Complete the following table of values relating the price of fuel (in dollars per gallon) and the number of miles Reggie can drive on $15 worth of gas.

Price of fuel (in dollars per gallon)	Number of miles Reggie can drive on $5 worth of gas
3.299	
3.449	
3.579	
3.839	

    e. How many miles can Reggie drive on $15 worth of gas if gas costs p dollars per gallon?

40. Alejandra bought a new house recently and is planning to install landscaping next week. In her initial budget, she set aside $200 to purchase ¾-inch gravel to cover part of her yard. This size gravel covers approximately 130 square feet per ton. She has four different options for this size gravel depending on the quality of the gravel. In addition to the charge for gravel the company charges $59 for delivery.

¾-inch gravel type	Cost per ton (in dollars)
Grade A	$31.50
Grade B	$26.50
Grade C	$24.50
Grade D	$22.50

    a. i. How many tons of grade B gravel can Alejandra purchase on her budget if she must pay the $59 delivery charge?
    ii. How many tons of grade D gravel can Alejandra purchase on her budget if she must pay the $59 delivery charge?
    b. How many square feet of her yard can she cover with grade C gravel? Explain how you found your answer.
    c. How many square feet of her yard can she cover with grade A gravel? Grade B gravel? Grade D gravel?

41. Jessie does a lot of traveling for business. When he travels he likes to go for a run in the morning to keep fit. Jessie has noticed that the elevation of the city he is visiting impacts how long he is able to spend running since there is less oxygen available at higher elevations. The graphs below provide information about Jessie's exercise routines recorded over many business trips.

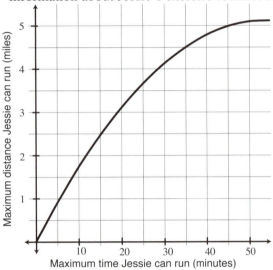

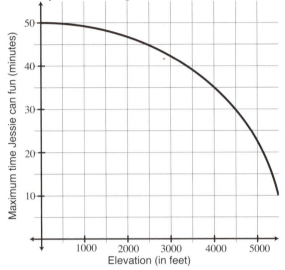

   a. Describe how to determine the maximum distance Jessie can run if he is visiting a city with an elevation of 4000 feet.
   b. Explain how you can determine the elevation of the city Jessie is visiting if he expects his maximum running distance to be 5 miles.

42. A ball is thrown into a lake, creating a circular ripple whose radius travels outward at a speed of 7 cm per second. The goal of this problem is to express the area of the circle as a function of the number of seconds that have elapsed since the ball hit the lake.
   a. Identify the quantities in the situation whose values vary and state what units you'll use to measure each of these quantities.
   b. Identify the quantities in the situation whose values are fixed and state what units you will use to measure each of these quantities.
   c. Draw a diagram of the situation and label the relevant quantities in the situation.
   d. As the amount of time t in seconds since the ball hit the lake increases over each of the given time periods, how does the radius r of the ripple (in centimeters) change?
      i. from $t = 0$ to 3 seconds
      ii. from $t = 4$ to 6 seconds
      iii. from $t = 6$ to 6.5 seconds
   e. Define a function g that defines the radius r of the ripple in terms of the time t in seconds since the ball hit the water.
   f. Define a functions f that determines the area of the ripple A in terms of the time t in seconds since the ball hit the water.
   g. Simplify your set of functions defined in part (e & f) in order to define a function h that expresses the area of the circle as a function of the time since the ball struck the water. Define all variables, including their units of measurement.
   h. Suppose $h(t)$ represents the area of the circle (measured in square centimeters) when t seconds have passed since the ball hit the water. Describe the meaning of $h(2.3)$ without performing any calculations. Then calculate and interpret the meaning of the value of this expression.

43. When hiring a contractor to add insulation to your attic, he provides you with the following tables that specify how input and output values are related for two functions, *f* and *g*.

Function *f*		Function *g*	
Number of bags of insulation applied, *n*	Depth of insulation (inches), *f*(*n*)	Depth of insulation (inches), *d*	Estimated annual heating/cooling costs (dollars), *g*(*d*)
11	2	15	$940
20	3.6	13	$1,000
28	5	11	$1,100
40	7.2	9	$1,250
50	9	7	$1,500
61	11	5	$1,900
70	12.6	4	$2,750
78	14	3	$3,600

a. Describe the meaning of $g(11)$.
b. Does the expression $f(g(11))$ have a real-world meaning in this context? If so, find the value and explain its meaning. If not, explain your reasoning.
c. Does the expression $g(f(50))$ have a real world-meaning in this context? If so, find the value and explain its meaning. If not, explain your reasoning.
d. Solve the equation $g(f(n)) = 1900$ for *n*. Then explain what your solution represents in this context.
e. Function *T* is defined as follows: $T(n) = g(f(n))$. Use the diagram below to describe what each of the parts (*A*, *B*, *C*, *D*, and *E*) of the definition represents.

$$\overbrace{\underbrace{T(n) = g}_{A}\underbrace{(f}_{C}(n))}$$

$$T(n) = g(f(n))$$

44. Suppose you have a coupon for $10 off a purchase of $100 or more at Better Buys.
a. Define a function *C* that determines the final cost of a purchase that totals *x* dollars before the coupon is applied. Explain what the expression $C(215.83)$ represents.
b. What are the domain and range of the function *C*?
c. This weekend only Better Buys has a customer appreciation sale that offers 5% off all purchases. Define a function *T* that determines the final cost of a customer's purchase after the sale if the original price of a purchase is *p* dollars. (Do not include the coupon in this function rule.)
d. Suppose Better Buys will allow you to use both the sale and the coupon together. If you buy a television set that is priced at $1979.99, describe what the expression $C(T(1979.99))$ represents and determine its value.

45. A spherical ball is inflated so that its radius increases by 2.4 cm every second. (Note the volume of a sphere is given by: $V = \frac{4}{3}\pi r^3$).
a. Draw a diagram of the situation and label the relevant quantities.
b. Define the following functions:
   i. *f* that defines the radius *r* (measured in centimeters) as a function of the number of seconds elapsed since beginning to inflate the ball, *t*.

(problem continues on next page)

    ii. *g* that defines the volume of the ball, *V*, (measured in cubic centimeters) as a function of the radius of the ball, *r* (measured in centimeters).

    iii. *h* that defines the volume of the ball, *V*, (measured in cubic centimeters) as a function of the number of seconds elapsed since beginning to inflate the ball, *t*.

  c. As the number of seconds since the ball began inflating increases from 4 seconds to 4.7 seconds what is the change in the volume of the ball?

  d. As the number of seconds since the ball began inflating increases from 5 seconds to 5.7 seconds what is the change in the volume of the ball?

  e. In this situation, the radius of the spherical ball was increasing at a constant rate with respect to the number of seconds elapsed since the ball began inflating. How would the pump have to be operating in order for this to occur? Justify your answer. (Hint: consider how the volume of the sphere changes for equal changes in the amount of time since the ball began to inflate.)

46. A landscaper plants circular flowerbeds in square plots of land of various sizes. Once the flowerbed is built she fills in the uncovered land with sod (non-shaded region in the figure below). As a result, the amount of sod needed is determined by the size of the flowerbed, which depends on the size of the square plot of land she uses.

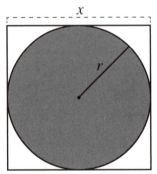

  a. Define the following functions:

    i. a function *f* that expresses the length of the side of the square, *x*, as a function of the radius of the circular flowerbed, *r*.

    ii. a function *h* that expresses the area of the square plot, *S*, in terms of the radius of the circular flowerbed, *r*.

    iii. a function *k* that expresses the area of the circular flowerbed, *A*, in terms of the radius of the circular flowerbed, *r*.

    iv. a function *g* that expresses the area of the grass needed to fill in around the circular flowerbed, *A*, as a function of the radius *r* of the circle.

  b. If the radius of the circular flowerbed is 9.5 feet, how much sod will the landscaper need to fill in the area around the flower bed?

  c. Define a function *t* that expresses the radius of the circular flowerbed, *r*, as a function of the length of one side of the square plot.

  d. What is the meaning of $g(t(x))$?

  e. Define a function *s* such that $s(x) = g(t(x))$. *Represent* (do not calculate) the amount of sod the landscaper needs to fill the area around the flowerbed for a square plot that is 12.3 feet by 12.3 feet.

47. An oil tanker crashed into a reef off of the coast of Alaska, grounding the tanker and punching a hole in its hull. The tanker's oil radiated out from the tanker in a circular pattern. You are an engineer for the oil company and your task is to monitor the spill.
    a. What are the varying quantities in this situation? Define variables to represent the values of the varying quantities. Be sure to include the units of measure when defining the variables.
    b. If the oil tanker was 105 feet from the spill's outer edge exactly 24 hours after the spill, how much area (in square miles) did the spill cover at that time.
    c. News reporters want to know how much oil has spilled at any given time. The only information available to you is satellite photos that include the spill's radius (in feet). You also know that 7.5 gallons of oil creates 4 square feet of spill.
       i. Define a function g that expresses the number of gallons of oil spilled, x, in terms of the area of the oil spill, A, (in square feet).
       ii. Define a function, f, that expresses the number of gallons of oil spilled in terms of the spill's radius (in miles).
    d. The spills radius at the end of 24 hours since the spill began was 105 feet. It is now 36 hours after the spill began and the spill's radius is 129 feet. Reporters want to know if oil is spilling faster or slower now than before. Use your function from part (c(ii)) to answer their question.
    e. Use your function from part (c(ii)) to represent the increase in the number of gallons spilled as the radius of the oil spill increases from 121 feet to 152 feet.

48. Assume that $f(x) = 2x+1$ and $g(x) = x^2 + 2x + 1$.
    a. Fill in the first two tables and use these two tables to complete the third table.

x	$f(x)$
0	
1	
1.5	
2	
5	

x	$g(x)$

x	$g(f(x))$
0	
1	
1.5	
2	
5	

    b. Describe, in words, how you used the first two tables to determine the values of the third table.

49. Use the tables below to answer the following questions

x	$f(x)$
−4	16
−3	13
−2	6
−1	0
0	−4
1	−7
2	−9
3	−10

x	$g(x)$
−8	−20
−6	−17
−4	−10
−2	−4
0	−1
2	2
4	6
6	11

    a. Evaluate the following expressions:
       i. $g(f(-2))$ ii. $f(g(0))$ iii. $g(g(4))$ iv. $f(g(2))$
       v. $g(f(-1))$ vi. $f(f(0))$
    b. Solve the equation $g(f(x)) = -10$ for x.
    c. Solve the equation $f(g(x)) = 16$ for x.

50. Use the words *input* and *output* to express the meaning of the following expressions. Assume that g and h are functions.
    a. $h(g(2))$
    b. $g(g(3))$
    c. $g(1+h(7.1))$

51. The functions f, g, and h (given below) are used to define the functions s, r, and v. Re-write the definitions of s, r, and v so that their definitions do not involve function composition.
    $f(x)=\frac{2x+1}{x-3}$ $g(x)=x^2-7$ $h(x)=-3x+2$
    a. $s(x)=f(g(x))$ b. $r(x)=h(h(x))$ c. $v(x)=g(h(x))$

52. The functions g, h, and k are defined below. Use these functions to answer the questions that follow.

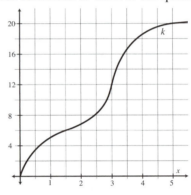

$$g(x)=x+5$$

x	$h(x)$
-1	-13.4
0	-7.3
2	-4.4
4	3
5	6.8
6	15
8	22

    a. $k(g(0))$ b. $h(k(1.5))$ c. $g(h(2))$ d. $k(h(4))$
    e. Solve the equation $h(g(x))=22$ for x.
    f. Solve the equation $g(k(x))=17$ for x.

53. Use the graphs below to answer the questions that follow

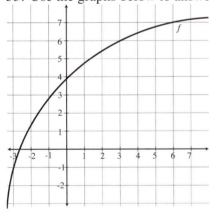
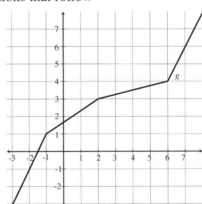

    a. Approximate the value of each of the following expressions.
        i. $f(f(3))$
        ii. $g(f(6))$
        iii. $g(g(-2))$
        iv. $g(f(-3))$
    b. Find the value of x that satisfies each of the following equations.
        i. $f(g(x))=6$
        ii. $g(f(x))=4$

54. Functions g and r are defined by their graphs below.

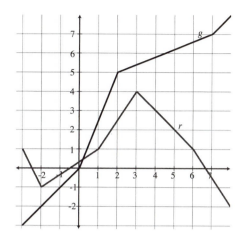

    a. Determine the values of each of the following expressions:
       i. $r(g(2))$
       ii. $g(r(1))$
       iii. $r(r(6))$
       iv. $g(r(-2))$
    b. How does the output $g(r(x))$ vary as x varies from 1 to 3?

55. For each of the below functions, redefine that function in terms of two new functions, f and g, using function composition and function arithmetic.
    a. $h(x) = 3(x - 1) + 5$
    b. $m(x) = (x + 4)^2$
    c. $k(\underline{x}) = (x + 2)^2 + 3(x + 2) + 1$
    d. $j(x) = \sqrt{x - 1}$
    e. $p(x) = \frac{500}{100 - x^2}$

IVb: Function Composition

56. Janet works 36 hours a week between two part-time jobs during the semester: residence hall front desk assistant ($7.50 per hour) and supermarket cashier ($9.25 per hour).
    a. Let r represent the number of hours Janet works in a given week as a residence hall front desk assistant. Write a formula to calculate the value of s, the number of hours Janet works as a supermarket cashier the same week.

    b. Define a function f that calculates the amount of money Janet makes for working r hours as a resident hall front desk assistant and then define a function g that calculates the amount of money Janet makes as a supermarket cashier working r *hours as a residence hall front desk assistant.*
    c. Define a function T that calculates the total amount of money Janet makes in a given week if she is working both jobs and works r hours at the residence hall front desk.
    d. The supermarket requires Janet to work between 15 and 30 hours each week. What is the domain of function T? What is the range of function T?

57. A spherical bubble inflates so that its volume increases by a constant rate of 120 cubic centimeters per second.
    a. Define a function, f, that expresses the radius of the bubble, r, (in centimeters) as a function of the bubble's volume, V, (in cubic centimeters). (Note the volume of a sphere is given by: $V = \frac{4}{3}\pi r^3$).
    b. Define a function g that expresses the bubble's volume, V, (in cubic centimeters) as a function of the number of seconds, t, since the bubble began to inflate.
    c. Use your functions from (a) and (b) to define a function h that expresses the bubble's radius, r (in centimeters) as a function of the number of seconds, t, since the bubble began to inflate.
    d. Use function notation to represent the change in the bubble's radius as the number of seconds since the ball began to inflate increases from 5 to 5.3 seconds.

58. A farmer decides to build a fence to enclose a rectangular field in which he will plant a crop. He has 1000 feet of fence to use and his goal is to maximize the area of his field.
    a. Draw a diagram that shows the quantities in this situation. Label the changing quantities with variables, and explain what each variable represents in terms of the situation. (e.g., define each quantity and its unit of measure).
    b. What are the dimensions of the enclosed field if one side must be 200 feet? What are the dimensions of the enclosed field if one side must be 352.41 feet? Justify your answer.
    c. Determine a formula that defines the length of the field l in terms of the field's width w given that the total amount of fence is 1000 feet.
    d. Describe how the length of the field changes as the width of the field increases.
    e. Determine a formula that relates the length and width of the field to the total area of the enclosed field.
    f. Using parts (c) and (e), define a function f that expresses the area of the field (measured in square feet) as a function of the length of the side of the field l (measured in feet).
    g. Create a graph that represents the area of the enclosed field in terms of the length of the side of the field. Describe how the area of the enclosed field changes as the length of the side of the field increases.
    h. Based on your graph, what is the maximum area of the enclosed field? What are the dimensions of the enclosed field that create this maximum area?

59. Brian decided to open a wildlife park and needs to build a square pen for giraffes. Brian has not decided how many giraffes he wants to acquire, but he knows that he should allow 10,000 square feet of grazing ground for each giraffe. Define a function f that expresses the number of feet of fencing Brian needs for his square pen as a function of the number of giraffes, n, that he acquires for his park.

60. The concentration of a certain pollutant, c, (in parts per million) depends on the population of a city (in thousands of people). The city of Edmonton is concerned about the concentration of this pollutant. Let p be the population of Edmonton (in thousands of people) and let t be the number of years since 2010. Suppose $c = 2.54p + 98.55$ and $t = 0.68p - 18.91$.
    a. Write a function g that determines the concentration of a certain pollutant, c, (in parts per million) in terms of the number of years, t, since 2010.
    b. Suppose that this pollutant is considered harmful when the concentration is more than 410 parts per million. Will the pollutant be considered harmful in the year 2070?
    c. Determine in what year the pollutant will be considered harmful.

61. You are planning a trip to Japan where the currency is the yen. On your way to Japan you will stop in Italy where the currency is the euro. You know that you will need to convert your US dollars to euros before your trip and know that the number of euros is 0.78 times as large the number of dollars. You also know the number of yen is 103 times as large as number of dollars.
    a. Write a formula that expresses the number of yen, y, you will have if you begin with d dollars.
    b. Write a formula that expresses the number of euro, e, you will have if you begin with d dollars
    c. You know that you will not be converting from dollars to yen because you will first stop in Italy. So it would be more helpful to know the conversion rate between euros and yen. Write a function g that expresses the number of yen you will have, y, in terms of the number of euros you convert.

62. The length of a steel bar changes as the temperature rises. Consider a 10 meter steel bar that is placed outside in the morning when the temperature is 20 degrees Celsius. Let l represent the length of a steel bar in meters. Let t be the temperature of the steel bar in degrees Celsius. And let n be the number of minutes since you placed the bar outside. You are given the following relationships: $l = 0.00013(t-20) + 10$ and $n = 12t - 240$. Write a function f that gives the length of the bar, l, (in meters) in terms of the number of minutes elapsed, n, since the bar was placed outside.

63. Let $n = 240s + 123$ and let $t = 0.14s^2$. Write a function h that gives n in terms of t.

64. Let $160 = 2t + 4m$ and let $p = \frac{1}{4}tm^2$. Write a function k that gives p in terms of t.

V. FUNCTION INVERSES (Text: S6)

65. For each of the following functions:
    i. describe the function process (For example: the input is increased by a factor of 4/3 to obtain the output)
    ii. the process that undoes the process of each of the given functions. (For example: If f is a process that increases its input by 7, the process that undoes f will decrease its input by 7).
    iii. algebraically define a process that undoes the process of the given function – that is, define the inverse function for each of the given functions.
    a. $f(x) = \frac{7}{5}x$
    b. $g(x) = 10 + x$
    c. $h(x) = \frac{x}{12}$

66. a. Define algebraically a process f that multiplies its input by 3, and then decreases this result by 1.
    b. Describe the process that undoes the process of f described in part (a).
    c. Algebraically define the process f^{-1} that undoes the process of f that is described in part (a).

67. Given the function f defined by $f(x) = 3x + 2$,
    a. Define f^{-1}, the function that undoes the process of f.
    b. Show that $f\left(f^{-1}(x)\right) = x$ for the function f that is given above.
    c. Show that $f^{-1}\left(f(x)\right) = x$ for the function f that is given above.
    d. What do you conclude about the relationship between f and f^{-1}

68. Given $g(x)=14x-9$, what are the value(s) of the input x when $g(x)$ equal to 109?

69. Given $h(x)=\frac{1}{6}x^2+7$, what are the value(s) of the input x when $h(x)$ equal to 31?

70. Given that $h(x)=4x$ determines the perimeter of a square given a square's side length x
    a. Algebraically define h^{-1} and describe its input quantity, output quantity, and process.
    b. Describe the input quantity, function process, and output quantity for $h\left(h^{-1}(x)\right)$.

71. Assume that g and z are functions that have inverse functions. Use the words *input* and *output* to express the meaning of the following expressions.
    a. $g^{-1}(21)$
    b. $z^{-1}(32)=43$

72. Find the inverse of the following functions. Then, determine if the inverse that you found is a function. Justify your answer.
    a. $g(x)=2x+4$
    b. $h(x)=2x^3-6$
    c. $k(x)=\frac{1}{2}x^2+12$
    d. $m(x)=\dfrac{2}{x+3}$

73. After Julia had driven for half an hour, she was 155 miles from Denver. After driving 2 hours, she was 260 miles from Denver. Assume that Julia drove at a constant speed. Let f be a function that gives Julia's distance in miles from Denver after having driven for t hours.
    a. Determine a rule for the function f.
    b. Interpret $f^{-1}(500)$. Calculate $f^{-1}(500)$.
    c. Determine a rule for f^{-1}.
    d. Construct a graph for $f^{-1}(d)$.

74. The functions f and g are defined in the following table. Based on this table, answer the questions below.

x	-3	-2	-1	0	1	2	3
$f(x)$	9	2	6	-4	-5	-8	-9
$g(x)$	3	0	3	2	-3	-1	-5

    a. Is f^{-1} a function? Explain why or why not.
    b. Is g^{-1} a function? Explain why or why not.
    c. Evaluate the following expressions:
       i. $f^{-1}(-9)$ ii. $g(3)$ iii. $f(g(2))$ iv. $f^{-1}(2)$
       v. $f^{-1}(g(3))$ vi. $g(f^{-1}(-4))$

75. The graphical representation of the functions *g* and *h* is given below. Use this information to answer the following questions.

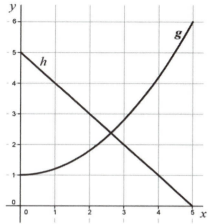

a. As *x* increases from 0 to 5, how does *g(x)* change? Be specific.
b. As *x* increases from 0 to 5, how does *h(x)* change?
c. Use the graphs above to evaluate the following:

    i. $g^{-1}(4)$

    ii. $g^{-1}(h(4))$

    iii. $g(h^{-1}(4))$

76. Given the functions $f(y) = \frac{3}{2}y + 21$ and $g(x) = 2x + 1$, answer the following questions.

a. Determine f^{-1} and explain how f^{-1} relates to f in terms of the input values and output values of each function.
b. Find $g^{-1}(g(2))$. Find $g^{-1}(g(3.5))$.
c. What pattern do you observe about $g^{-1}(g(x))$? Explain this pattern.
d. Determine the composition function $f(g(x))$ and explain, using input-output language, the meaning of the function $f(g(x))$ relative to the functions f and g.
e. Considering the description of the function $f(g(x))$ given in part (d), describe the inverse of this function. Justify whether $f^{-1}(g^{-1}(x))$ or $g^{-1}(f^{-1}(z))$ is the proper notation for representing this function's inverse, and then determine the rule of the chosen function.
f. What are the domain and range of the function f?

77. Apply the ideas of function inverse and function composition to answer each of the following.
a. Define the area of a square *A* in terms of its perimeter *p*.
b. Define the diameter of a circle *d* in terms of its circumference *C*.

78. Apply the ideas of function inverse and/or function composition to answer each of the following.
a. Define the perimeter of a square *p* in terms of its area *A*.
b. If the radius of a circle is growing at 8 cm per second, define the area of a circle *A* in terms of the amount of time *t* since the radius was 0 cm and started growing.
c. Define the diameter of a circle *d* in terms of its area *A*.

VI. DIFFERENCE QUOTIENT (Text: S7)

79. Consider the function $f(x) = x^2 - 6x + 10$ that represents the altitude of a US Air Force test plane (in thousands of feet) during a recent test flight, as a function of elapsed time (in minutes) since being released from its airborne launcher.
    a. Find the average rate of change of the plane's altitude with respect to time as the time varies from $x = 2$ minutes to $x = 2.1$ minutes. Show your work.
    b. What is the meaning of the *average rate of change* you determined in part (a) in this context?
    c. Explain the meaning of the expression $f(k + 0.1)$.
    d. Explain the meaning of the expression $f(k + 0.1) - f(k)$.
    e. Suppose you have been given the graph of $y = f(x)$ below and a specific value of x, say $x = k$.
       Show $f(k + 0.1) - f(k)$ on the following graph.

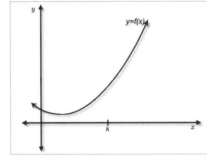

    f. What does the expression $\dfrac{f(k+0.1) - f(k)}{(k+0.1) - k}$ represent?

For problems 80-84, determine the difference quotient (e.g., the average rate of change over an input interval of length h) for the given functions. Be sure to simplify your answer.

80. $f(x) = 12x + 6.5$

81. $f(x) = 9.7$

82. $f(x) = 6x^2 + 7x - 11$

83. $f(x) = 3x^3 - 9$

84. $f(x) = \dfrac{1}{2x}$

Percentages

A *percentage* refers to a type of measurement where the measurement unit is a specific value of some quantity. To be exact, **1% is $\frac{1}{100}$ of the value of the reference quantity**. Imagine that on one day during a recent trip you drove 340 miles. If we use this value as our reference, then "1% of the distance you drove on that day" corresponds to $\frac{1}{100}$ of 340 miles.

340 miles

1% of 340 miles, or
$\frac{1}{100}$ times as large as 340 miles

*1. a. What distance corresponds to 1% of the distance you drove on that day?

   b. To find the distance in miles in part (a), we can multiply 340 miles by what number?

   c. If you had driven 1% of the 340 miles, what percent of the total distance do you still have left to drive? What is this distance in miles?

   d. To find the distance in miles in part (c), we can multiply 340 miles by what number?

*2. a. What distance corresponds to 36% of the distance you drove on that day?

   b. To find the distance in miles in part (a), we can multiply 340 miles by what number?

   c. If you had driven 36% of the 340 miles, what percent of the total distance do you still have left to drive? What is this distance in miles?

   d. To find the distance in miles in part (c), we can multiply 340 miles by what number?

3. A local cable company recently raised their rates. Their premium entertainment package used to cost $85 per month but now it costs $99 per month. Imagine using the old price as a "measuring stick" to measure the new price as shown.

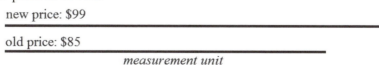

new price: $99

old price: $85

measurement unit

   a. How many times as large is the new price compared with the old price? What does this mean in terms of a measurement process?

b. Suppose we want to fill in the following blank: "The new monthly price is ____% of the old monthly price." To answer this question, what do we need to think of as our "measuring stick" (unit)? How is this different from the measurement in part (a)?

c. Fill in the blank in part (b).

*4. In order to make room for next year's models, a car dealership reduced the price of all new cars on their lot. One car that used to cost $24,995 now costs $22,355. Suppose we use the old price as a "measuring stick" to measure the new price as shown.

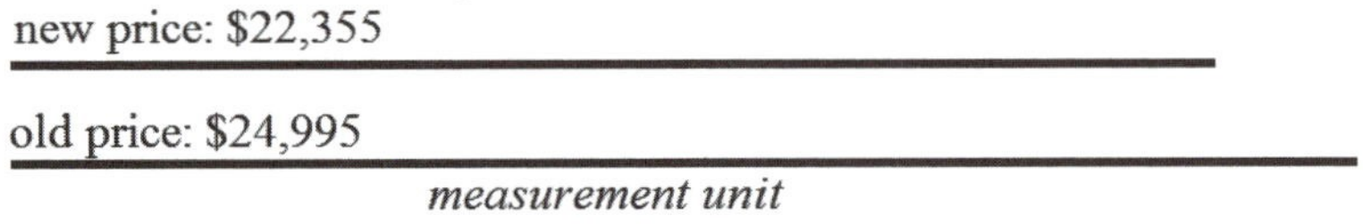

new price: $22,355
_____

old price: $24,995
_____
measurement unit

a. How many times as large is the new price compared with the old price? What does this mean in terms of a measurement process?

b. Suppose we want to fill in the following blank: "The new price is ____% of the old price." To answer this question, what do we need to think of as our "measuring stick" (unit)? How is this different from the measurement in part (a)?

c. Fill in the blank in part (b).

In Exercises #1-4, we saw how a percentage is a measurement based on using $\frac{1}{100}$ of the quantity's value as the "measuring stick" (unit). For example, paying an 8% tax means the tax we pay is 8 times as large as $\frac{1}{100}$ of the purchase price. [*Note: We usually condense this statement to say that the tax we pay is $\frac{8}{100}$ of the purchase price.*] But this means that tax charged (in dollars) is different for different purchase prices even though the tax percentage (8%) remaining constant.

Similarly, a sale where all prices are discounted by 30% uses $\frac{1}{100}$ of each individual price as a "measuring stick" (unit) so that the discount (in dollars) is different for each original price even though the discount percentage (30%) remains constant. See the following diagram that illustrates this idea.

retail price is $75
_____

discount is $22.50

retail price is $49   retail price is $26
_____   _____

discount is $14.70   discount is $7.80

the discount is 30% in each case
(30 times as large as $\frac{1}{100}$ of the retail price, or $\frac{30}{100}$ of the retail price)

5. For each price and discount given, find the value of the discount in dollars.
    a. price was $114, discount is 20%   b. price was $28, discount is 45%

*6. For each price and price increase given, find the value of the increase in dollars.
    a. price was $52, increase is 35%   b. price was $199, increase is 215%

Percent Change

Percent change refers to thinking about the *difference* between two values of a quantity, measured using $\frac{1}{100}$ of one of the values as the measurement unit. For example, if the price of an item was $50 and increased to $59, then we can say the following.

-   The change in price is $9.

-   The percent change from the old price to the new price is $\frac{9}{50}$, or $\frac{18}{100}$, or 18% of the old price.

-   The new price is $59, which is 118% of the old price of $50.

Note that it's common (and good practice) to describe a *decrease* as a negative change. For example, if the original price of an item was $80 and decreased to $62, then we can say the following.

-   The change in price is –$18.

-   The percent change from the original price to the new price is $-\frac{18}{80}$, or –0.225, or $-\frac{22.5}{100}$, or –22.5% of the original price.

-   The new price is $62, which is 87.5% of the original price.

*7. Suppose you walk into a store that is having a sale on many of its items. For each sale described, do the following.
    i)   State the percent change in the price.
    ii) State the number we can multiply the original price by to find the sale price.
    iii) Find the sale price in dollars.

    a. original price: $150, sale: 40% off   b. original price: $19, sale: 10% off

c. original price: $915.99, sale: 15% off   d. original price: $22.99, sale: 12.5% off

8. Suppose a store needs to raise prices on some of its items. For each price increase described, do the following.
    i)   State the percent change in the price.
    ii) State the number we can multiply the original price by to find the new price.
    iii) Find the new price in dollars.

    a. original price: $52, increase: 10%   b. original price: $13, increase: 50%

    c. original price: $14.99, increase: 2.3%   d. original price: $1,499.99, increase: 100%

*9. In 2010 an investment was worth $12,000. By 2015, the investment had lost 20% of its value (a percent change of –20%).
    a. What was the value of the investment by 2015?

    b. Suppose the investor wants the value of the investment to return to its 2010 value within the next 3 years. She wants to know by what percent her investment must change in order to regain its former value. Determine the percent change necessary for her investment to return to its previous value.

10. Suppose you bought a pair of jeans on sale for $18, which was discounted 25% from their normal retail price. Suppose you want to know the normal retail price. What is the normal retail price?

11. A new company expanded rapidly and doubled the number of its employees every month over the last year.
    a. Fill in the blank: The number of employees one month was _____% of the number of employees in the previous month.

    b. What was the percent change in the number of employees from one month to the next?

*12. A motion picture company released the new trailer for an upcoming movie yesterday. Today, its website received 5 times as many visitors as it did yesterday.
   a. What was the percent change in the number of visitors from yesterday to today?

   b. Fill in the blank: The number of visitors today was _____% of the number of visitors yesterday.

Meaning of Exponents

Exponents are used to represent how many factors are included in a product. For example, each of the following products are shown rewritten in exponential notation along with their equivalent real-number value (decimal notation).

product notation	$4 \cdot 4 \cdot 4 \cdot 4 \cdot 4$	$2 \cdot 2 \cdot 2 \cdot 2 \cdot 5 \cdot 5 \cdot 5$	$7 \cdot 6 \cdot 4 \cdot 6 \cdot 7$
exponential notation	4^5	$2^4 \cdot 5^3$	$4 \cdot 6^2 \cdot 7^2$
decimal notation	1,024	2,000	7,056

*13. Imagine that Jerry weighed 180 pounds and his weight increased by 5% each year for 4 years.
   a. Complete the missing entries in the table.

	product notation	exponential notation	decimal notation
starting weight (lbs)	180	180	180
weight after 1 year (lbs)	180(1.05)		189
weight after 2 years (lbs)		$180(1.05)^2$	
weight after 3 years (lbs)	180(1.05)(1.05)(1.05)		
weight after 4 years (lbs)			218.79

   b. During this four-year period, did Jerry's weight change by the same number of pounds per year? Why or why not?

   c. If *n* represents the number of years since Jerry weighed 180 pounds and *w* represents his weight in pounds, write a formula that represents *w* in terms of *n*.

d. After 4 years, what was Jerry's ending weight gain in pounds? What is this change as a percentage of his starting weight?

e. After 4 years, Jerry's weight is what percent of his starting weight?

*14. How does your answer to Exercise #13c change if Jerry's initial weight was 210 pounds instead of 180 pounds?

*15. How does your answer to Exercise #13c change if Jerry's weight decreased by 5% each year instead of increased by 5% each year?

*16. Simplify the following expressions and write the final result using only positive exponents.

a. $\left(2x^2\right)^3\left(3x\right)^2$

b. $\left(a^2b\right)^4\left(ab^3\right)^5$

c. $\left(2c^3d^2\right)^6\left(2cd\right)^3$

d. $\dfrac{14a^5p^2}{7ap^2}$

e. $\dfrac{x^4y^{10}z^3}{x^2y^5z^5}$

f. $\dfrac{5(x+1)^3(x-7)^4}{15(x+1)^2(x-7)^9}$

When examining situations with varying quantities, we typically want to identify pairs of quantities that change together (co-vary). When we identify such a pair, our next goal is to understand *how* their values change together and, if possible, to identify patterns in the relationship. In other words, when studying co-varying relationships we want to identify what stays the same while the quantities change.

In a previous module we closely examined *linear functions*. A linear relationship exists when two quantities' values change together with a *constant rate of change*. In such a relationship the two quantities' values are free to vary, but they must do so in such a way that the ratio of their changes $\frac{\Delta y}{\Delta x}$ is a constant. But not all co-varying relationships exhibit a constant rate of change. In this investigation we will encounter examples of a different type of relationship and identify what remains the same as two quantities co-vary.

For Exercises #1-5, use the following information. When a piece of new technology is invented and sold, its value decreases over time because it wears down and because even newer, better technology is developed to replace it. Suppose two electronic devices have the same resale value right now and that their resale values are expected to change according to the following patterns.

- **Product #1: iTech Device** The current resale value is $300. The resale value is expected to decrease 30% per year for the next several years.
- **Product #2: Dynasystems Device** The current resale value is $300. The resale value is expected to decrease $45 per year for the next several years.

*1. a. Find the expected resale value (in dollars) for both devices 1 year from now and describe how you determined each value.

b. Without performing additional calculations, predict which device will have a greater resale value 6 years from now.

c. Complete the given table showing the expected resale value for each device over the next 6 years.

Δt	Years from now, t	Expected resale value for the iTech Device, s (dollars)	Δs		Δt	Years from now, t	Expected resale value for the Dynasystems Device, n (dollars)	Δn
	0	300				0	300	
	1					1		
	2					2		
	3					3		
	4					4		
	5					5		
	6					6		

d. Check your prediction from part (b) by using the table in part (c).

e. The table on the left demonstrates how the expected resale value for the iTech Device co-varies with the number of years elapsed from now. What stays the same as these two quantities' values change together?

f. The table on the right demonstrates how the expected resale value for the Dynasystems Device co-varies with the number of years from now. What stays the same as these two quantities' values change together?

*2. a. Define function formulas to model the expected resale value in dollars for each device as functions of the number of years from now, t.

iTech Device: $s = f(t)$; where $f(t) =$ _____

Dynasystems Device: $n = g(t)$; where $g(t) =$ _____

b. Explain what each part of the formulas represents relative to the context we are exploring.

3. Find and interpret the following using your functions from Exercise #2.
  a. $f(4)$ b. $g(3)$

  c. $g(6) - g(2)$ d. $f^{-1}(147)$

*4. In Exercise #1, the table on the left has a column showing values of $s = f(t)$ (the expected resale value for the iTech Device in dollars) as well as a column showing values of Δs (the change in the expected resale value for the iTech Device in dollars as the number of years from now changes by 1).

a. What are the ratios of consecutive entries in the column for s? That is, what are the values of $\frac{f(1)}{f(0)}$, $\frac{f(2)}{f(1)}$, $\frac{f(3)}{f(2)}$, and so on?

b. How does your answer to part (a) relate to the original problem context? In other words, what do these ratios tell us about the resale value of the iTech Device?

c. What are the ratios of values of Δs compared to the values of s at the beginning of each interval?

d. How does your answer to part (c) relate to the original problem context? In other words, what do these ratios tell us about the resale value of the iTech Device?

The following are graphs of functions f and g that model the expected resale values for each device over the next 6 years. On each graph we have plotted the points that correspond to the ordered pairs you should have calculated in Exercise #1.

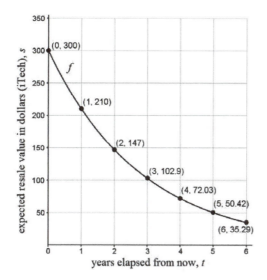
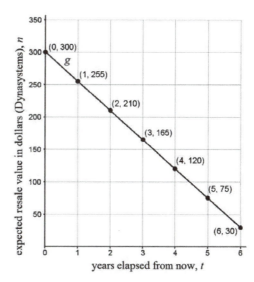

To help focus attention on the values of s and n (the expected resale values of each device in dollars) represented in our previous tables, we have drawn vertical line segments whose lengths represent values of s and n respectively.

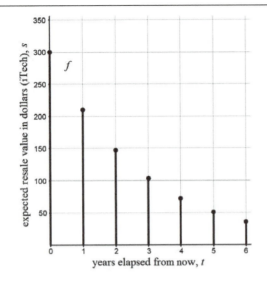
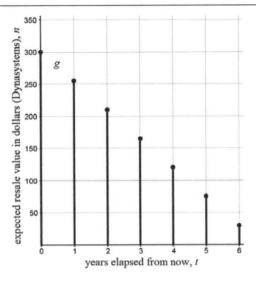

*5. a. Explain how we can "see" on the graph the fact that the expected resale value of the iTech Device at some moment is 70% of its expected resale value one year earlier.

   b. Explain how we can "see" on the graph the fact that the expected resale value of the iTech Device changes by –30% each year.

   c. Explain how we can "see" on the graphs that only one of the two devices has an expected resale value that changes at a constant rate of change with respect to the number of elapsed years.

6. Suppose another piece of technology (the iTech2 Device) has a current resale value of $400, and that this value is expected to change by –50% per year. Let function *h* model its expected resale value in dollars *t* years from now.

   a. *Without performing any calculations*, plot the approximate values for the ordered pairs $(1, h(1))$, $(2, h(2))$, $(3, h(3))$, $(4, h(4))$, $(5, h(5))$, and $(6, h(6))$.

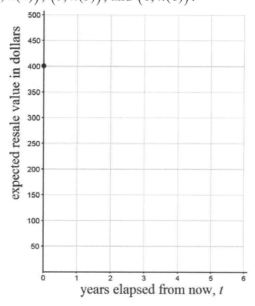

   b. Explain how you were able to estimate the values for the ordered pairs in part (a).

   c. Suppose that another device (TecTone) has a current resale value of $500 and its resale value is expected to decrease by 20% each year. Let function *j* model its expected resale value *t* years from now. Plot the point (0, 500) on the graph in part (a), then *without performing any calculations*, plot the approximate values for the ordered pairs $(0, j(0))$, $(1, j(1))$, $(2, j(2))$, $(3, j(3))$, $(4, j(4))$, $(5, j(5))$, and $(6, j(6))$.

   d. Write formulas to represent functions *h* and *j*.

Exponential Functions

When two quantities change together such that, for equal changes in one quantity's values, the second quantity's values have a ***constant percent change***, then we say that the relationship is an ***exponential function***.

---

*7. In Exercises #1-6 you explored several examples of relationships that can be modeled with an exponential function. Using your work with these examples, explain why exponential functions *do not* have a constant rate of change of one quantity with respect to the other quantity.

*8. From about 2000 to 2006, home prices in the United States increased dramatically prior to the real estate "crash" that began around 2007-2008. Suppose that from 2000 to 2006 the value of a specific house in California increased by 25% per year.
   a. Complete the following table and draw a graph modeling the value of this house in dollars *n* years since the beginning of 2000.

years since the beginning of 2000	home price (in dollars)
0	210,000
1	
2	
3	
4	
5	
6	

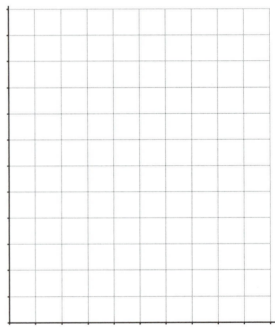

   b. By what number can we multiply the value of the home at one moment in time to find its value one year later?

   c. Fill in the blank: The value of the home at one moment in time is _____% of its value one year earlier.

   d. Explain how we can "see" the value of the multiplier you identified in part (b) in the graph from part (a).

*1. The function $f(x) = 2^x$ is an example of an exponential function, and some people call this a *doubling function*.

    a. If we allow x to vary, and it increases by 1, what happens to the output value of the function?

    b. When x varies by 1,
        i. the new output value is _____% of the old output value.

        ii. What is the percent change in the output value?

    c. If we allow x to vary, and it increases by 3, what happens to the output value of the function?

    d. If we allow x to vary, and it increases by 8, what happens to the output value of the function?

In the previous investigation we only explored exponential functions with a limited domain (such as $0 \le x \le 6$). However, without a context to restrict the value of x, the function $f(x) = 2^x$ can accept any real number as its input and produces the following graph.

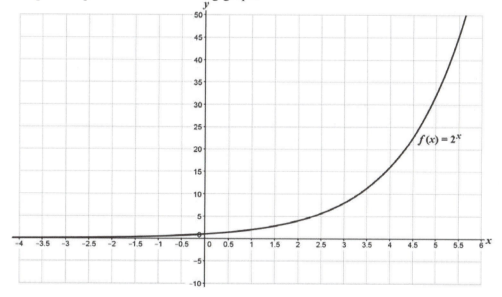

2.   a. Determine the value of $f(0)$.

    b. As x increases away from $x = 0$, how do the function values change? What happens as x gets very, very large?

c. As x decreases away from $x = 0$, how do the function values change? What happens as x gets very, very small (becomes a large magnitude negative number)?

*3. View the Module 4 Investigation 3 PowerPoint animation titled **Exercise 3 Animation**.
   a. What idea or ideas are being demonstrated in this animation?

b. How is this image different from thinking about the function using a table of values like the one shown?

x	0	1	2	3	4	5
$f(x)$	1	2	4	8	16	32

*4. In Exercises #1-3 we looked at a specific example of a doubling function $f(x) = 2^x$. Give three more examples of exponential functions whose values double whenever x increases by 1.

5. Consider the function $g(x) = 3^x$.
   a. On the graph shown prior to Exercise #2, add the graph of g.

   b. Compare the behavior of functions f and g. How are they alike? How do they differ?

   c. For function g, whenever x increases by 1,
      i. the new function value is _____% of the old function value.

      ii. What is the percent change in the output value?

   d. For function g, whenever x increases by 2, what happens to the function's output value?

© 2015 Carlson, Oehrtman, and Moore

6. Consider the function $h(x) = 6\left(\frac{2}{3}\right)^x$.

    a. On the graph shown prior to Exercise #2, add the graph of h.

    b. As x increases away from $x = 0$, how do the function values change? What happens as x gets very, very large?

(1-Unit) Growth Factor

For an exponential function, the ratio of output values is always the same for equal-sized intervals of the domain. When x increases from x_1 to x_2, this ratio $\frac{f(x_2)}{f(x_1)}$ is called a **growth factor**. If the interval is 1-unit wide (that is, if the change in x from x_1 to x_2 is 1), then the ratio is the **1-unit growth factor**.

The growth factor is a useful number because it tells us how many times as large $f(x_2)$ is compared to $f(x_1)$, and thus can be used to determine $f(x_2)$ if we know $f(x_1)$.

Note that if this factor is between 0 and 1 it is often called a **decay factor** because a multiplier between 0 and 1 means that the function values are decreasing as x increases.

---

*7. The following table shows values for an exponential function f.

x	$f(x)$	$\Delta f(x)$
1.24	3.1000	
2.24	4.3400	
3.24	6.0760	
4.24	8.5064	

a. What are the ratios for the function's output values when x varies by 1? That is, what is the value of $\frac{f(2.24)}{f(1.24)}$, $\frac{f(3.24)}{f(2.24)}$, and $\frac{f(4.24)}{f(3.24)}$?

b. Why does the answer to part (a) tell us the function's (1-unit) growth or decay factor?

c. Complete the third column in the table showing the differences in the output values when x varies by 1.

d. What are the values of the ratios for each change in the output value compared to the value of the function at the beginning of the interval? What is the percent change in the output value when x increases by 1 (this is the 1-unit percent change)?

---

*8. View the Module 4 Investigation 3 PowerPoint animation titled **Exercise 8 Animation**. What idea or ideas are being demonstrated in this animation?

In Exercises #9-12, do the following.
  a. Find the ratio of output values that correspond to increases of 1 in the input value in order to determine the growth or decay factor.
  b. Determine the 1-unit percent change by comparing the change in the output values to the function value at the beginning of a 1-unit interval for x.
  c. Identify or determine the value of the function when $x = 0$.
  d. Use the information from parts (a) through (c) to define a function formula for the relationship.

*9.

x	0	1	2	3
$f(x)$	16	4	1	0.25

  a.

  b.

  c.

  d.

10.

x	1	2	3	8
$g(x)$	260	299	343.85	691.605

  a.

  b.

  c.

  d.

11.

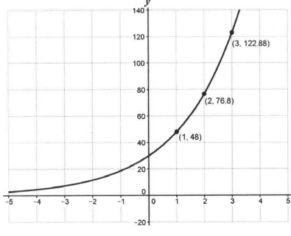

  a.

  b.

  c.

  d.

*12.

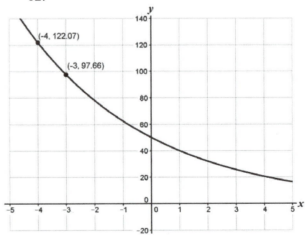

  a.

  b.

  c.

  d.

13. Let $f(x) = 34(1.19)^x$.

    a. What does the number "34" represent for this function?

    b. What does the number "1.19" represent for this function?

    c. Fill in the blank: Whenever x increases by 1, the new output value is _____% of the old output value.

    d. What is the 1-unit percent change and what does it tell us?

*14. Let $g(p) = 1.578(0.68)^p$.

    a. What does the number "1.578" represent for this function?

    b. What does the number "0.68" represent for this function?

    c. Fill in the blank: Whenever p increases by 1, the new output value is _____% of the old output value.

    d. What is the 1-unit percent change and what does it tell us?

*15. An investment of $3300 increases by 4.7% each month.

    a. What is the 1-month percent change in the investment value?

    b. Fill in the blank: When the time elapsed since the investment was made increases by 1 month, the new value of the investment is _____% of the old value of the investment.

c. What is the 1-month growth or decay factor and what does this value tell us about the situation?

d. Write a function formula to model the value of the investment (in dollars) in terms of the time elapsed since the investment was made (in months).

16. After having 1.4 million people at the start of 2010, the population of a city has been decreasing by 2.1% per year.
    a. What is the 1-year percent change in the city's population?

    b. Fill in the blank: When the time elapsed since the beginning of 2010 increases by 1 month, the new population is _____% of the old population.

    c. What is the 1-year growth or decay factor and what does this value tell us about the situation?

    d. Write a function formula to model the city's population (in millions) in terms of the time elapsed since the beginning of 2010 (in years).

*17. Some of your classmates made the following claim. "The functions $f(x) = x^2$ and $g(x) = x^3$ are some other examples of exponential functions."
    a. Are your classmates correct? Justify your answer using what you've learned so far in this module.

    b. If you agree with your classmates, then come up with at least two more examples of exponential functions. If you disagree with your classmates, then give a possible reason for why this might be a common mistake for students.

*1. Suppose that you know (4, 38) and (7,50) are two ordered pairs for an *exponential growth function*. Your classmate noticed that the difference in the outputs is 12 as the inputs change by 3, and created the following table showing some additional ordered pairs.

x	y
4	38
5	**42**
6	**46**
7	50

Explain how you know that the two additional ordered pairs (bolded) cannot be accurate even though we don't have the function formula or the graph for this relationship.

2. Urban planners noticed that the population of a certain county doubled over 20 years (from 1990 to 2010), following a pattern of exponential growth. The population of the county was 756,000 people at the end of 1990.

  a. What was the population of the county at the end of 2010?

  b. If the growth pattern were to continue for another 20 years, what is the projected population at the end of 2030?

  c. Which quantities are changing in this situation? Which quantities are not changing?

  d. If we assume that this population growth will continue into the far future, complete the following table.

Number of years since 1990	Population	Number of times larger the current population is than the population at the end of 1990
0		
20		
40		
60		
80		

  e. Define a function *f* that expresses the population of the county, *P,* after a number of 20-year periods, *n*, that have elapsed since the end of 1990.

3. Since 20-year changes in time are quite long, we might want to know how much the population changes over shorter periods of time.
   a. Given that the population grew exponentially, which of the following best describes what happens every 10 years?
      i. Every 10 years, the population increases by a constant amount. That is, a constant amount is added to the population at the end of 1990 to get the population at the end of 2000, and the same amount is added to get the population at the end of 2010.
      ii. Every 10 years, the population increases by a constant factor. That is, this constant factor is multiplied by the population at the end of 1990 to get the population at the end of 2000, and this constant factor is multiplied by the population at the end of 2000 to get the population at the end of 2010.

   b. Given that the population increases exponentially and doubles over 20 years,
      i. What is the 10-year growth factor?

Number of years since 1990	Population
0	
1	
2	
3	
13	
16	
20	

      ii. What is the 1-year growth factor?

      iii. What is the 2-year growth factor? What is the 3-year growth factor? The 7-year growth factor?

      iv. Complete the given table.

   c. Define a function g that expresses the population of the county, P, as a function of the number of years t since the end of 1990.

   d. Define a function h that expresses the population of the county c months since the end of 1990.

*4. A farm in Canada had 218 alpacas on January 1, 2004. After 8 years the alpaca population decreased to 187. Assume the number of alpacas decays exponentially.
   a. What is the 8-year growth/decay factor?

   b. Fill in the blank: The number of alpaca on the farm on January 1, 2012 was _____% of the number of alpaca on the farm on January 1, 2004.

   c. What is the 8-year percent change?

   d. Assuming the alpaca population continues to be modeled by the same exponential model, how many alpacas can we expect to be on the farm on January 1, 2020?

   e. Since 8-year changes in time are quite long, we might want to know how the population of alpacas changes over shorter periods of time. What is the 1-year growth/decay factor?

   f. What is the 1-year percent change?

   g. Define a function f that relates the number of alpacas on the Canadian farm t years from January 1, 2004 (Assume the alpaca population continues to change by the same decay factor each year.)

*5. Assume the number of alpacas continues to change by the same decay factor each year as defined in Exercise #3. Use your calculator determine the following;
   a. After how many years will there be 109 alpacas remaining on the farm, assuming the herd started with 218 alpacas?

   b. After how many total years will there be 55 alpacas?

c. After how many years will a herd of 120 alpacas decrease to 60 alpacas?

*6. Assume the number of alpacas continues to change by the same decay factor each year as defined in Problem #3.
   a. What is the 1-month growth/decay factor?

   b. Define a function *g* that relates the number of alpacas on the Canadian farm *k* months from January 1, 2004 (Assume the alpaca population continues to change by the same decay factor each year).

   c. What is the 2-month growth/decay factor?

7. After taking medicine, your body begins to break it down and remove it according to a pattern of exponential decay. Suppose you take 500 mg of Ibuprofen and that your body removes 85% of the Ibuprofen every 5 hours.
   a. After 5 hours, what percent of the 500 mg dose remains in your body? How much medication is this?

   b. Define a function *f* that expresses the amount of Ibuprofen, *B*, present after *n* 5-hour time intervals since taking the medicine.

   c. Define a function *g* that expresses the amount of Ibuprofen, *B*, present *t* hours after taking the medicine.

   d. How much Ibuprofen remains in your body 3 hours after taking the medicine? After 12 hours?

   e. How does the *change* in the amount of Ibuprofen remaining change as the number of hours elapsed increases?

For Exercises #1-2, determine the specified growth or decay factor, percent change, and initial value for each of the following exponential functions.

1. $f(x) = 9.5(1.24)^x$

    a. 1/2-unit Growth Factor:

    b. 2-unit Growth Factor:

    c. 2-unit Percent Change:

    d. Initial Value:

*2. $g(x) = 0.46(0.874)^{4x}$

    a. 1/4-unit Decay Factor: $.402 = 0.46(0.874)^1$

    b. 1-unit Decay Factor: $.268 = 0.46(0.874)^4$

    c. 5-unit percent change: $\cancel{.031}\ \dfrac{.268 - .031}{2.68} = 12\%$

    d. Initial Value: 0.46

In Exercises #3-4, determine the specified growth or decay factor, percent change, initial value, and function for each of the following tables modeled by exponential functions.

3.

x	0	2	4	6
$f(x)$	16	10.24	6.554	4.194

    a. 2-unit Decay Factor:

    b. 1-unit Percent Change:

    c. 3-unit Decay Factor:

    d. ½-unit Decay Factor:

    e. Initial Value:

    f. Function:

*4.

x	1	4	7	10
$g(x)$	260	278.2	297.674	318.511

    a. 3-unit Growth Factor: $1.06 = 278/260$

    b. 6-unit Percent Change: $1.14 = 297.674/260$

    c. 1-unit Growth Factor: $1.02\ (278/260)^{1/3}$

    d. ¼-unit Growth Factor: $1.01 = (278/260)^{1/12}$

    e. Initial Value: 260

    f. Function: $260(1.07)^{1/4 x}$

In Exercises #5-6, determine the specified growth or decay factor, percent change, initial value, and function formula for the exponential function with the given graph.

*5. Use the graph below to determine the following values.

    a. 2-unit Growth/Decay Factor:

    b. 6-unit Growth/Decay Factor:

    c. 1-unit Percent Change:

    d. ½-unit Growth/Decay Factor:

    e. 3-unit Growth/Decay Factor:

    f. Initial Value:

    g. Function:

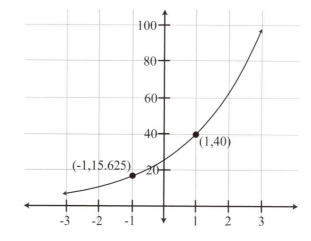

6. Use the graph below to determine the following values.
   a. 3-unit Growth/Decay Factor:

   b. 6-unit Growth/Decay Factor:

   c. 1-unit Percent Change:

   d. ½-unit Growth/Decay Factor:

   e. 5-unit Growth/Decay Factor:

   f. Initial Value:

   g. Function:

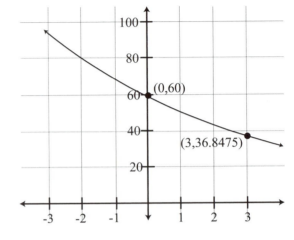

In Exercises #7-9, determine the growth or decay factor, percent change, initial value and function for each exponential situation.

7. An investment of $6300 increases by 7.3% each year.
   a. 2-year Growth Factor: b.   1/2-year Percent Change:

   c. Initial Value: d.   Function:

*8. After having 1.97 million people in 2010, the population of a city has been decreasing by 4.2% every three years.
   a. 1-year Decay Factor: b.   ¼ -year Percent Change:

   c. Initial Value: d.   Function:

9. The amount of caffeine in your body decreases by 27% every 4 hours.
   a. 24-hour Decay Factor: b.   24-hour Percent Change:

   c. 1-hour Decay Factor: d.   1-hour Percent Change:

When you invest money the bank will pay interest on the money in the account. The interest can be paid in many different ways such as once a year, twice a year, once a month, etc. By convention, banks calculate the interest rate per compounding period by dividing the advertised annual interest rate (often called the Annual Percentage Rate or APR) by the number of compounding periods in one year.

*1. Suppose you are going to invest $1000. You have three choices on how to invest your money and in all three situations the annual interest rate (APR) is 8%.
    a. The first account advertises, "8% compounded annually". This means that at the end of each year 8% of the current balance is added to the value of the account.
        i. What is the interest rate per compounding period?

        ii. Complete the following table.

8% Compounded Annually	
number of years since investment was made	value of the investment (in dollars)
0	
1	
2	
t	

        iii. What is the annual growth factor for this account? Interpret the annual growth factor in the context of this problem.

    b. The second account advertises "8% compounded semiannually". This means that interest is added two times per year (every six months).
        i. What is the interest rate per compounding period?

        ii. Complete the following table.

8% Compounded Semiannually	
number of years since investment was made	value of the investment (in dollars)
0	
0.5	
1	
2	
t	

        iii. What is the six-month growth factor for this account?

    iv. What is the annual growth factor for this account? Interpret the annual growth factor in the context of this problem.

    v. Use the annual growth factor computed in part (iv) to determine the actual percentage change of the account over a 1-year time period. This percentage is often referenced as the annual percentage yield (APY).

c. The third account advertises "8% compounded daily". This means that interest is added 365 times per year (every day).

    i. What is the interest rate per compounding period?

    ii. Complete the following table.

8% Compounded Daily	
number of years since investment was made	value of the investment (in dollars)
0	
1/365	
2/365	
1	
2	
t	

    iii. What is the daily growth factor for this account?

    iv. What is the annual growth factor for this account? Interpret the annual growth factor in the context of this problem.

    v. Use the annual growth factor computed in part (iv) to determine the annual percentage yield (APY).

d. Which of the three accounts would you chose for your investment? Justify your answer.

*2. A bank offers an annual interest rate of r compounded n times per year, where r is the APR expressed as a decimal. Let a be the initial value of the investment. Define a function g that determines the value of the investment B in terms of the number of years since the investment was made, t.

3. Bank USA is offering a 9% annual rate compounded monthly and Southwest Investment Bank is offering an annual rate of 8.5% compounded daily. Which bank would you choose for your investment?

*4. You should have noticed that the techniques in this investigation for finding growth factors related to different units of time are not the same as the techniques used in previous investigations. This is because bank policy differs from our goals in creating partial growth factors. Look at the following comparison.

Non-Financial Contexts	**Compound Interest**
The population of a certain country is 26.4 million people and is increasing by 4% per year. Annual growth factor: 1.04 Annual percent change: 4% Monthly growth factor: $\qquad (1.04)^{1/12} \approx 1.003274$ Monthly percent change: 0.3274% When the monthly growth factor is applied 12 times it produces the annual growth factor. $\qquad [(1.04)^{1/12}]^{12} = 1.04$	*$1200 is deposited into an account with an APR of 4%.* If interest is compounded once per year, the annual growth factor is: 1.04 The annual percent change is: 4% If interest is compounded once per month, the annual growth factor is: $1 + \frac{0.04}{12} = 1.00\overline{3}$ The monthly percent change is: $0.\overline{3}\%$ If interest is compounded once per month, and we do this for one year, the annual growth factor is different than when interest is compounded only once per year. $\qquad (1.00\overline{3})^{12} \approx 1.04074$

a. In non-financial contexts, we know the actual annual growth factor based on data and we cannot change the long-term growth pattern of the function relationship. How does this play a role in the methods we use for computing growth factors for different-sized changes in the input quantity's value?

b. Why don't banks have the same restrictions as those described in part (a) and how might this impact their techniques?

*5. A deposit of $5,000 is made into an account paying an annual interest rate of 5%. Determine the amount in the account 10 years after the investment was made if the interest is compounded:
a. Annually

b. Monthly

c. Weekly

d. Daily

*1. Suppose $1000 is invested into an account with an annual percent rate (APR) of 8%. Given the number of compounding periods determine the value of the account at the end of the fifth year. Also determine the annual percent yield (APY) for the given compounding period. (Round to at least 3 decimal digits.)

	APY	Difference in the value of the APY	Value of Investment after 5 years	Difference in the value of the investment
Compounded Yearly				
Compounded Quarterly				
Compounded Monthly				
Compounded Daily				
Compounded Hourly				
Compounded Every Minute				

2. How is the number of compounding periods per year related to the annual growth factor and the APY? Why do you think this is happening?

*3. a. What do you anticipate the APY will be if the APR is 8% and the investment is compounded every second, every $1/100^{th}$ of a second?

   b. What do you predict is the largest annual growth factor possible given an APR of 8%? Interpret the meaning of this value in the context of this situation.

*4. Consider investment APRs of 5%, 6%, 7%, and 8%.
   a. Complete the following table by predicting the largest annual growth factor that corresponds to those APRs. (Round to at least 6 decimal digits)

APR	5%	6%	7%	8%
Largest Annual Growth Factor, g				
$g^{1/r}$ (where r is the APR as a decimal)				

   b. What do you notice about the values in the second row of the table?

An interesting mathematical idea is that no matter what the APR is for a given investment, the largest annual growth factor for that investment is related to the number 2.71828... This irrational number is referred to as e. Confirm this by using your calculator to find the value of e.

*5. The value of e is equal to $g^{1/r}$ where g is the largest annual growth factor and r is the decimal value of the given APR. Undo this process to find an equation for g in terms of e and r.

*6. Suppose an investment of $1000 was made in four different banks at the given APRs. Complete the table by finding the largest annual growth factor for each APR. This value should be a decimal. Since this decimal is a rounded value (and therefore not exact), complete the second row by writing an expression that determines the exact value of the largest annual growth factor. Finally, write a formula that can be used to find the maximum value of the investment after t years (the value of the investment assuming the largest annual growth factor).

APR	5%	6%	7%	8%
Largest Annual Growth Factor (decimal form)				
Exact Representation of the Largest Annual Growth Factor				
Maximum Value of Investment After t Years				

*7. Define a function f that gives the value of an investment, B, in terms of the number of years since the initial investment was made, t. Let a represent the value of the initial investment and let r represent the decimal form of the APR.

Another way to think about the largest annual growth factor is to say that it is the growth factor that corresponds to compounding as many times as possible in a year. In other words, if the investment were to be compounded more often than daily, more often than hourly, more often than every second, then we would say the investment is being compounded continuously.

8. Suppose that $1500 is invested into an account with an advertised APR of 7% compounded continuously.
   a. Determine the value of the investment after 6 years.

   b. Determine the amount of time required for the value of the investment to double.

   c. Determine the value of the account after 6 years had the investment been compounded quarterly instead of continuously.

Up until this point we have been unable to algebraically solve equations where the unknown variable is in the exponent. We utilized our graphing calculators in order to solve these equations. In this investigation we will learn how to solve these equations algebraically.

*1. a. Without a calculator approximate the solution to the following equations.

     i. $2^x = 10$

     ii. $3^x = 10$

   b.   Describe the process you used to determine your solutions in part (a).

Determining the solutions of the equations in Exercise #1 involved undoing the process of exponentiation. Instead of raising a base to a power we determined the power to which a number (the base) is raised to obtain some number. The function that undoes the process of exponentiation is called the logarithmic function. The input of the logarithmic function is the output of the exponential function (the result of raising some base to a power) and the output of the logarithmic function is the input of the exponential function (the exponent or power to which the base is raised).

The standard way of writing the logarithmic function is to write "log" with a subscript specifying the base to which x is raised to obtain the number that is input to the function. As an example, $\log_2(10) = x$ is read "log base 2 of 10 equals x". We can determine the value of x by considering the power that 2 is raised to in order to obtain a value of 10".

2. Rewrite each of the following equations in logarithmic form (if possible). If it is not possible, say why.

   a. $4^x = 64$ b. $5^x = \frac{1}{125}$ c. $2^x = -32$

We can rewrite any exponential equation in logarithmic form. Note that the input to the logarithmic function can only be non-negative real numbers.

Logarithmic Function

For $x > 0$ and $b > 0$, with $b \neq 1$,

     $y = \log_b x$ is equivalent to $b^y = x$. $x - b^y$

The function $f(x) = \log_b x$ is the logarithmic function with base b.

---

*3. Rewrite the following exponential equations in logarithmic form.

   a. $y = 4^x$ b. $b = 1.5(5)^a$ c. $m = 2^{4t}$ d. $q = 3(5)^{2k}$

*4. Without using your calculator, determine/estimate the value of the unknown.

a. $\log_2 4 = y$ b. $\log_9\left(\frac{1}{81}\right) = t$ c. $\log_3(-2) = k$ d. $\log_5 10 = s$

e. $\log_7 18 = p$ f. $3\log_4 22 = b$ g. $\log_5 5 = d$ h. $-2\log_3 27 = z$

*5. Logarithmic functions are the inverses of exponential functions. That is, both functions show the same relationship between two quantities, but the input and output quantities are switched.

a. The graph of $f(x) = \log_4(x)$ is given. Based on your understanding of logarithms, explain why f is an increasing function and also why y increases less and less for equal changes in x as x increases.

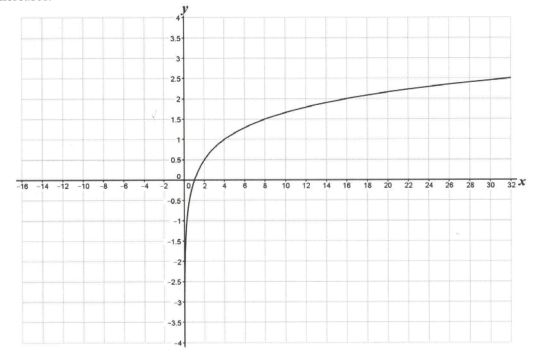

b. On the axes in part (a), draw the graph of $g(x) = \log_2(x)$.

c. On the axes in part (a), draw the graph of $h(x) = \log_{0.5}(x)$.

Recall that in Module 4 Investigation 7 we defined the irrational number $e = 2.718281828\ldots$ A logarithm with base e or $\log_e c = x$ is commonly called the natural logarithm and is expressed $\ln c = x$.

Note: Your calculator has two logarithm buttons, log and ln. Even though we can use any base (greater than zero) in a logarithmic function, most calculators only evaluate logarithms for two bases: base 10 and base e. The log x button evaluates $\log_{10} x$ and is referred to as the **Common Log**. The ln x button evaluates $\log_e x$ and is referred to as the **Natural Log** of x.

6. Convert the following to exponential form and evaluate/estimate the value of the unknown. Check your answer using your calculator.

    a. $\log\frac{1}{100} = x$ b. $\ln e^2 = k$ c. $\ln 7 = t$ d. $\log 1000 = s$

Recall the Properties of Exponents:
Property of Exponents #1: $b^x \cdot b^y = b^{x+y}$

Property of Exponents #2: $\dfrac{b^x}{b^y} = b^{x-y}$ for $b \neq 0$.

Property of Exponents #3: $\left(a^x\right)^y = a^{xy}$

Because the result from evaluating a logarithmic expression (and the output of a logarithmic function) represents an exponent, the properties of exponents also apply to logarithms, with the rules of exponents expressed using logarithms and logarithmic form.

Here are three properties of logarithms used to re-write logarithm expressions as a single logarithm.
Property of Logarithms #1: $\log_b(m \cdot n) = \log_b m + \log_b n$ for any $b > 0$, $b \neq 1$, and $m, n > 0$

Property of Logarithms #2: $\log_b\left(\dfrac{m}{n}\right) = \log_b m - \log_b n$ for any $b > 0$, $b \neq 1$, and $m, n > 0$

Property of Logarithms #3: $\log_b\left(m^c\right) = c \cdot \log_b(m)$ for any $b > 0$, $b \neq 1$, and $m, n > 0$

The first property of logarithms states $\log_b(m \cdot n) = \log_b m + \log_b n$. For example, $\log_2 4 + \log_2 8$ asks to find the power to which 2 is raised to in order to get a result of 4, and the power 2 is raised to in order to get a result of 8, and then add those two powers together $(2 + 3 = 5)$. We could have instead considered $\log_2(4 \cdot 8) = \log_2(32)$ and determined directly that 5 is the power 2 must be raised to in order to get a result of 32.

*7. Use the properties of logarithms to simplify the following expressions.

    a. $\ln(x) + \ln(x)$ b. $\log_3(5) + \log_3(2)$ c. $\log_4(16) - \log_4(4)$

$\ln(x \cdot x)$

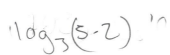

d. $\log_3(2) + \log_5(3)$ e. $\log_3(9) + \log_3(4) - \log_3(6)$ f. $\log_7(49) - \log_7(-3)$

*8. Solve each of the following for x.
    a. $\log_5(30x^2) = 3$ b. $\log(1.5x) = 0$ c. $\log_3 x + \log_3(2x) = 3$

9. Let $f(x) = 10^x$ and $g(x) = \log(x)$.
    a. Complete the following tables.

x	$f(x)$
-2	
-1	
0	
1	
2	

x	$g(x)$
0.01	
0.1	
1	
10	
100	

    b. Evaluate the following expressions:
       i. $g(f(-2))$
       ii. $g(f(0))$
       iii. $f(g(1))$
       iv. $f(g(10))$
       v. $f(g(x))$
       vi. $g(f(x))$

    c. What do you notice about the relationship between the functions f and g?

*1. Solve each of the following equations for x. Find the exact answer and then use your calculator to approximate the answer to the nearest thousandth (3 decimal places).

    a. $4 = 3^x$

    b. $22.4 = 17.5(3.4)^x$

2.   a. Why is it mathematically valid to "take the natural log" of both sides of an equation? (In other words, how do you know the two sides of the equation are still equivalent?)

    b. Is it mathematically valid to take the natural log of both sides of an equation if one or both sides of the equation has a value less than or equal to zero? Explain.

*3. In 2009 the enrollment at Mainland High School was 1650 students. Administrators predict that the enrollment will increase by 2.2% each year

    a. Determine when the enrollment at Mainland High School is expected to reach 2600 students.

    b. Suppose that Mainland High School requires 5 teachers per 65 students. How many teachers are required if the student enrollment is 2600 students?

    c. Define a function f that gives the number of years since 2009 in terms of the number of teachers employed by Mainland High School

*4. Recall the alpaca problem from Module 4 Investigation 4 which stated, a farm in Canada had 218 alpacas on January 1, 2004. After 8 years the alpaca population decreased to 187. Assume the number of alpacas decays exponentially.

    a. Define a function f that expresses the number of alpacas on the farm in terms of the number of years since 2004.

b. Using the function defined in part (a), estimate the number of alpaca on the farm in 2013.

c. Determine the year when the number of alpaca on the farm was 165. Solve algebraically and then check your answer using a graph or by using your function defined in part (a).

5. In 1928 22 koala bears were introduced to Wallaby Island. In 1992, the population of koala bears on the island was about 2000. Assume the population of koala bears can be modeled by exponential growth.
   a. Determine the annual growth factor in this situation.

   b. Solving algebraically, determine in that year the population of koala bears on the island will be 4500.

6. You deposited $2,500 in a CD to help save for your college tuition. The annual interest rate of the CD is 4.5% compounded continuously.
   a. Define a function that gives the value of the CD in terms of the number of years that have elapsed since you opened the CD.

   b. How many years will have elapsed before the CD is worth $7,000?

7. Jasmine has $13,570 to invest in a savings account.
   a. Determine the number of years before her money doubles at an annual rate of 4% compounded continuously.

   b. Determine the number of years before her money doubles at an annual rate of 7% compounded monthly.

I. PERCENTAGES, PERCENT CHANGE, AND THE MEANING OF EXPONENTS (TEXT: S1)

1. Suppose that you worked a total of 450 minutes at your job yesterday.
   a. What amount of time corresponds to 78% of the total time you spent working yesterday?
   b. To find the amount of time in part (a) we can multiply 450 minutes by what number?
   c. If you had worked for 78% of the 450 minutes, what percent of the total time do you still have left to work? What is this time in minutes?
   d. To find the amount of time in part (c) we can multiply 450 minutes by what number?

2. Suppose that you fill a glass with 860 milliliters of water.
   a. What volume of water corresponds to 44% of 860 mL?
   b. To find the volume in part (a) we can multiply 860 mL by what number?
   c. If you drank 44% of the water you poured into the glass, what percent of the starting 860 mL remains? How much water is this in mL?
   d. To find the volume in part (c) we can multiply 860 mL by what number?

3. A local bicycle shop placed older model bikes on sale last weekend. Suppose a customer bought a bike on sale for $299 that had a normal retail price of $495. Imagine using the retail price as a "measuring stick" to measure the sale price as shown.

   sale price: $299
   _____

   retail price: $495
   _____
   measurement unit

   a. How many times as large is the sale price compared with the retail price? What does this mean in terms of a measurement process?
   b. Suppose we want to fill in the following blank: "The sale price is ____% of the retail price." To fill in this blank, what do we need to think of as our "measuring stick" (unit)? How is this different from the measurement in part (a)?
   c. Fill in the blank in part (b).

4. Use the context from Exercise #3, but this time suppose we want to use sale price as the "measuring stick" (unit) to measure the retail price.

   retail price: $495
   _____

   sale price: $299
   _____
   measurement unit

   a. How many times as large is the retail price compared with the sale price? What does this mean in terms of a measurement process?
   b. Suppose we want to fill in the following blank: "The retail price is ____% of the sale price." To fill in this blank, what do we need to think of as our "measuring stick" (unit)? How is this different from the measurement in part (a)?
   c. Fill in the blank in part (b).

5. Suppose that your doctor recommends that you increase or decrease your daily intake of certain vitamins and minerals. For each recommendation described, do the following.
    i)   State the percent change in daily intake.
    ii) State the number by which we can multiply the original daily intake to find the new daily intake.
    iii) Find the new recommended daily intake amount.
    a. original vitamin D intake: 3μg; increase: 60%
    b. original potassium intake: 1500mg; increase: 135%
    c. original zinc intake: 30mg; decrease: 45%
    d. original calcium intake: 820mg; increase: 25%

6. Suppose that a store is adjusting the prices of several items. For each price change, do the following.
    i)   State the percent change.
    ii) State the number by which we can multiply the old price to find the new price.
    iii) Find the new price.
    a. old price: $29; increase: 5%
    b. old price: $84; increase: 140%
    c. old price: $89.99; decrease: 34%
    d. old price: $6.49; decrease: 22%

7. A major record label has seen its annual operating profit decrease in recent years, likely because of greater accessibility of music online. In 2011, the label's operating profit was $135 million. By 2015, the label's operating profit had decreased by 43% (a percent change of –43%).
    a. What was the record label company's operating profit in 2015?
    b. Suppose the record label wants to increase its operating profit to $100 million by 2017. By what percent must the label's operating profit increase from its 2015 value to reach $100 million within the next two years?

8. During a recession, Tina had to take a 10% pay cut. Her original salary was $58,240.
    a. What was her salary after the pay cut?
    b. Once the recession was over, Tina's company wanted to increase her pay to her original salary. What percent change in her salary was required to return it to its original $58,240?

9. Complete the following table. Assume you start with 5 pennies on a table. On the first square of a chessboard you place triple the number of pennies you began with. You continue this pattern until you fill the chessboard.

Number of Pennies in a Square			
Square	Decimal Notation	Product Notation	Exponential Notation
1st	15	$5 \cdot 3$	
2nd	45		$5 \cdot 3^2$
3rd	135	$5 \cdot 3 \cdot 3 \cdot 3$	
4th	405		
5th	1215	$5 \cdot 3 \cdot 3 \cdot 3 \cdot 3 \cdot 3$	$5 \cdot 3^5$
6th	3645		
9th	98,415	$5 \cdot \underbrace{3 \cdot 3 \cdots 3}_{9 \text{ factors}}$	
14th	23914845		

10. For the following situations, define a function that describes the number of pennies on a given square of a chessboard. Let n represent the number of the square and $P = f(n)$ represent the number of pennies on the given square.
   a. Suppose you start with 4 pennies on a table. On the 1st square you double the initial amount of pennies. The number of pennies on any given square is double the number on the previous square.
   b. Suppose you start with 2 pennies on a table. On the 1st square you triple the initial amount of pennies. The number of pennies on any given square is triple the number on the previous square.
   c. Suppose you start with 6 pennies on a table. On the 1st square you quadruple the initial amount of pennies. The number of pennies on any given square is quadruple the number on the previous square.
   d. Suppose you start with 512 pennies on a table. On the 1st square you half the initial amount of pennies. The number of pennies on any given square is half the number on the previous square. (Assume that we can have fractions of pennies).

11. Last year, Jenny invested her birthday money in 3 different penny stocks. The following functions represent the daily value (rounded to the nearest cent) of each stock over the first 7 days after making the investment. For each investment,
    i) State the amount of the initial investment;
    ii) Describe how the value of the investment grew over the first 7 days since making the investment?
    iii) Determine the value of investment at the end of the 7th day. (Round your answers to the nearest penny.)
    a. $h(n) = 25 \cdot 1.45^n$
    b. $f(n) = 120 \cdot 3^n$
    c. $g(n) = 275 \cdot 0.9^n$

12. The rabbit population on a 10-acre wildlife preserve was 24 on January 1, 2011.
    a. Assuming the number of rabbits doubled each year, determine a function that gives the number of rabbits R in the preserve in terms of the number of years t elapsed since January 1, 2011.
    b. Using the function created in part (a), approximate the number of rabbits in the preserve on January 1, 2018.
    c. What is the percent change per year if the population of rabbits doubles each year?
    d. Assuming the number of rabbits instead doubled each *two years*, determine a function that gives the number of rabbits in the preserve in terms of the number of years elapsed since January 1, 2011.
    e. Assuming the number of rabbits instead *tripled* each *four years*, determine a function that gives the number of rabbits in the preserve in terms of the number of years elapsed since January 1, 2011.

13. A firework stand opened on June 28th. The number of customers at the stand over the first 5 days since it was opened is defined by the function, $N(x) = 7(2)^x$. At the end of July 2, the function was updated to $N(x) = 224(3)^x$, with x continuing to represent the number of days passed since the firework stand was opened. What implications can you draw from this information?

14. Simplify the following expressions and write the final result using positive exponents.
    a. $(4a)^3 \left(4^2 a^5\right)^2$
    b. $\left(xyz^3\right)^2 \left(\left(2x^3\right)^4 y\right)^{-4}$
    c. $\dfrac{(x+1)^2 (x-3)^3}{(x-3)^2 (x+1)^3}$

    d. $\dfrac{5a^0 b^0}{(5a)^{-5}}$
    e. $\dfrac{6\left(x^{-1} yz^3\right)^4}{\left(6x^{-1} yz\right)^{-2}}$
    f. $\dfrac{\left(3x^{-1} z^5\right)^{-2}}{\left(2x^5 y\right)^3}$

II. COMPARING LINEAR AND EXPONENTIAL BEHAVIOR (TEXT: S2)

15. Joni, a graduating biomedical engineer, was offered two positions, one with Company A, and the other with Company B. Company A offered her a starting salary of $66,000 with a 5% guaranteed raise at the end of each of the first five years. Company B offered her $75,000 as her starting salary with a guaranteed raise of $1,500 every year for the first 5 years. She likes both companies and believes she will continue with the company she selects for at least 5 years.
    a. What is the ratio of Joni's salary one year compared to her salary in the previous year for Company A? Describe how to interpret this ratio.
    b. Fill in the blank: If Joni works for Company A, her salary for any year is ____% of her salary for the previous year.
    c. Define a function f that expresses her salary at Company A in terms of the number of years n since she accepts the position.
    d. Define a function g that expresses her salary at Company B in terms of the number of years n since she accepts the position.

e. After how many years will Joni's salary at Company A overtake her salary at Company B?

f. Suppose that she will work for one of these companies for exactly 5 years. A classmate says she should choose Company A because by the time she leaves the company she will have a higher salary. Do you agree? Defend your reasoning.

g. Construct a graph of the two functions and explain the meaning of the intersection point in the context of this situation (Note that the graphs of the two functions will not be continuous. Before creating your graph think about the discrete instances when her salary changes and the fact that her salary remains constant during each year.)

h. Find and interpret the following using the functions created in parts (b) and (c).

    i. $f(8)$ ii. $g^{-1}(8200)$ iii. $g(16)$ iv. $f(20) - f(11)$

16. Suppose you are researching jobs in advertising.
    a. You are told that the salary for Job A increases exponentially. The salary for Year 1 is $29,000 while the salary for Year 2 will be $31,066.
       i. How many times as large is the salary in Year 2 compared to the salary in Year 1?
       ii. Fill in the blank: The salary in Year 2 is _____% of the salary in Year 1.
       iii. What is the percent change in salary from Year 1 to Year 2?
       iv. If the salary continues to increase by the same percent each year, what will the salary be in Year 5?
    b. You are told that the salary for Job B also increases exponentially. The salary for Year 1 is $27,500 while the salary for Year 2 will be $28,600.
       i. How many times as large is the salary in Year 2 compared to the salary in Year 1?
       ii. Fill in the blank: The salary in Year 2 is _____% of the salary in Year 1.
       iii. What is the percent change in salary from Year 1 to Year 2?
       iv. If the salary continues to increase by the same percent each year, what will the salary be in Year 5?
    c. Based on the information found in parts (a) and (b), which job would you take? Explain your reasoning.

17. A chemist monitored the mass of bacteria in a Petri dish after applying a chemical to kill the bacteria. This chemical causes the mass of bacteria of this type to decrease by 12% each hour that elapses after applying the chemical and the mass when applying the chemical was 203 micrograms.
    a. Fill in the blank. At any given time, the mass of the bacteria is _____% of the mass one hour earlier.
    b. What would the mass of bacteria be after 1 hour? After 2 hours? After 8 hours?
    c. Define a function that expresses the mass of bacteria B remaining in the Petri dish as a function of the number of hours t since applying the chemical.
    d. After how many hours since applying the chemical will the bacteria's mass be less than one microgram?
    e. Suppose that a different type of chemical was applied to another Petri dish of bacteria that caused the mass to decrease by 26% each hour after applying it. If the initial amount of bacteria in this dish was 230 micrograms, define a function which gives the mass of bacteria in this Petri dish as a function of the number of hours t that have elapsed since applying the chemical.

18. When a teacher asked her beginning algebra class to provide an example of an exponential function, over half of the students offered the function $f(x) = x^2$, indicating that the function is growing faster and faster so it is growing exponentially.
    a. What is your assessment of these students' answer?

b. Compare the growth patterns of $f(x) = x^2$ and $g(x) = 2^x$. Construct a table of values and create a graph of f and g on the same axes, then use the graphs and table values to describe and contrast their growth patterns.

c. Compare the growth patterns of $f(x) = x^2$ and $g(x) = 2^x$. Construct a table of values and create a graph of h and g on the same axes, then use the graphs and table values to describe and contrast their growth patterns.

19. Define a function that models each town's population growth in terms of the number of years since the town was established.
   a. Smallsville starts with 500 people and grows by 10 people per year.
   b. Growsville starts with 500 people and grows by 10% each year.
   c. Shrinktown starts with 500 people and declines by 10% of the population at the end of the previous year.
   d. Littletown starts with 500 people and declines by 10 people per year.

20. Each function below defines the population for a city in terms of the time t in years since the city was established. Write a sentence that describes the city's initial population and growth pattern.
   a. $f(t) = 2000(1.24)^t$
   b. $g(t) = 1500 + 20t$
   c. $h(t) = 4000(0.68)^t$
   d. $k(t) = 2500 - 40t$
   e. $f(t) = 1500(1.4)^{t/2}$

21. The US population was about 273.6 million in 1996. Since that time the population increased by approximately 1.1% each year.
   a. Define a function f that expresses the population P of the US in millions as a function of the number of years t since 1996.
   b. What was the approximate population of the US in 2010?
   c. Assuming the population continues to grow at 1.1% per year, in what year will the US population reach 400 million people?

22. A biologist counted 426 bees in a bee colony. His tracking of the colony revealed that the number of bees increased by 4% each month over the next year.
   a. Approximately how many bees were in the colony 3 months after the initial count?
   b. How many months passed before the number of bees reached 500?

23. A company purchases a new car for $25,000 for their employees to use. For accounting purposes, they decide to depreciate the value of the car by 14.5% each year.
   a. Using this method of depreciation, what is the value of the car after 2 years?
   b. What is the ratio of the car's value in one year compared to its value the previous year? Explain the meaning of the value of this ratio.
   c. Define a function that models the value of the car as a function of the number of years since the company purchased it.
   d. When will the value of the car be less than $1000?

24. A deposit of $5,100 is made into a savings account at an annual interest rate of 3.75%.
   a. Define a function that relates the amount of money in the account to the time t in years since the initial investment was made. (Assume no additional deposits or withdrawals are made.)
   b. How much money is in the account 4 years after the initial investment?
   c. How many years before the investment reaches a value of $10,000?

25. Simplify the following expressions.

    a. $\sqrt{25x^{12}}$
    b. $\sqrt{49x^7}$
    c. $\sqrt{90x^2y^6}$

    d. $\sqrt{120x^{13}y^5z}$
    e. $\sqrt[3]{27x^{12}}$
    f. $\sqrt[3]{24x^5y^9z^{10}}$

III. 1-UNIT GROWTH AND DECAY FACTORS, PERCENT CHANGE, AND INITIAL VALUES (TEXT: S2)

26. Determine the growth or decay factor, percent change, and initial value for each of the following exponential functions.

    a. $f(x) = 2^x$

        Initial Value:
        1-unit Growth Factor:
        1-unit Percent Change:

    b. $f(x) = (0.98)^x$

        Initial Value:
        1-unit Decay Factor:
        1-unit Percent Change:

    c. $f(x) = 0.56 \cdot (0.25)^x$

        Initial Value:
        1-unit Decay Factor:
        1-unit Percent Change:

    d. $f(x) = 3 \cdot (1.6)^x$

        Initial Value:
        1-unit Growth Factor:
        1-unit Percent Change:

27. The given tables represent patterns of exponential growth. Determine the initial value, 1-unit growth/decay factor, 1-unit percent change, and define a function to model the data in each table.

    a.

x	0	1	2	3
$f(x)$	512	384	288	216

    Initial Value:
    1-unit Decay Factor:
    1-unit Percent Change:
    Function Formula:

    b.

x	1	2	3
$g(x)$	11.2	15.68	21.952

    Initial Value:
    1-unit Growth Factor:
    1-unit Percent Change:
    Function Formula:

28. Determine the growth or decay factor, percent change, initial value, and function for each of the following graphs modeled by exponential functions. (*Exercise continues on the following page.*)

    a.

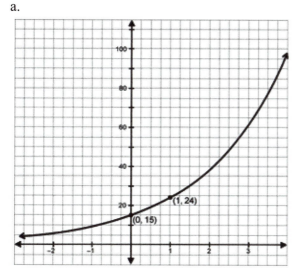

    b.

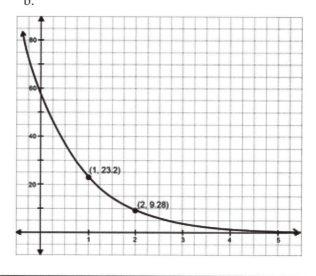

Initial value: Initial Value:
Growth Factor: Decay Factor:
Percent Change: Percent Change:
Function Formula: Function Formula:

29. An investment of $9000 decreased by 2.4% each month.
    a. Determine the initial value of the investment, the 1-month decay factor, and the 1-month percent change of the investment.
    b. Define a function *f* that determines the value of the investment in terms of the number of months since the initial investment was made.
    c. Determine the value of the investment 12 months after the money was invested.

30. The population of Canada in 1990 was 27,512,000 and in 2000 it was 30,689,000. Assume that Canada's population increased exponentially over this time.
    a. Determine the initial value, the 10-year growth factor, and the 10-year percent change of Canada's population.
    b. Define a function *f* that defines the population of Canada in terms of the number of *decades* (10-year periods) since 1990.
    c. Assuming Canada's growth rate remains consistent, predict the approximate population of Canada in 2030.

31. One year after an investment was made, the amount of money in the account was $1844.50. Two years after the investment was made, the amount of money in the account was $2001.28.
    a. Determine the initial value of the investment, the investment's 1-month growth factor, and the investment's 1-month percent change.
    b. Define a function *f* that defines the value of the investment in terms of the number of months since the investment was made.
    c. Determine when the value of the investment will reach $4000.

32. The mass of bacteria in a Petri dish was initially measured to be 14 micrograms and increased by 12% each hour over a period of 15 hours.
    a. Determine the 1-hour growth factor of the bacteria.
    b. By what total percent did the bacteria increase during the 15-hour time period?
    c. Determine how long it will take for the mass of bacteria to double.

IV. PARTIAL AND *N*-UNIT GROWTH FACTORS (TEXT: S3)

33. A certain strain of bacteria that is growing on your kitchen counter doubles every 5 minutes. Assume that you start with only one bacterium.
    a. What quantities are changing in this situation? What quantities are not changing?
    b. Define a function that gives the number of bacteria present after *n* 5-minute intervals.
    c. Represent the number of bacteria present after 30 minutes. How many bacteria are there at that time?
    d. How many bacteria are present after 1 minute? 2 minutes? 5 minutes?
    e. Determine the 1-minute growth factor. By what percent does the bacteria change every *minute*?
    f. Determine the 2-minute growth factor. By what percent does the bacteria change every two minutes?
    g. Define a function that gives the number of bacteria present after *t* minutes.
    h. Define a function that gives the number of bacteria present after *h* hours.

34. The number of deer in a park reserve was counted to be 27 on January 1, 2000. Six months later, the number of deer was counted to be 38.
    a. Determine the 6-month growth factor and 6-month percent change for the situation.
    b. Determine the 1-month growth factor and 1-month percent change for the situation.
    c. Define a function that relates the number deer in the park reserve since January 1st 2000 in terms of the number of months that have elapsed since January 1, 2000. (Assuming the number of deer continues to increase by the same percent each month.)
    d. Define a function that relates the number deer in the park reserve since January 1st 2000 in terms of the number of *6-month intervals* that have elapsed since January 1, 2000. (Assuming the number of deer continues to increase by the same percent each month.)
    e. Use one of your functions to determine when the number of deer will reach approximately 125.

35. The population of Egypt in 2002 was 73,312,600, and was 78,887,000 in 2006. Assume the population of Egypt grew exponentially over this period.
    a. Determine the 4-year growth factor and percent change.
    b. Determine the 1-year growth factor and percent change.
    c. Define a function that gives the population of Egypt in terms of the number of years that have elapsed since 2002.
    d. Re-write your function from part (c) so that it gives the population of Egypt in terms of the number of *decades* that have elapsed since 2002.
    e. Assuming Egypt's population continued to grow according to this model, how long will it take for population of Egypt to double?

36. An animal reserve in Arizona had 93 wild coyotes. Due to drought, there were only 61 coyotes after 3 months. Assume that the number of coyotes decreases (or decays) exponentially.
    a. Find the 3-month decay factor and percent change.
    b. Find the 1-month decay factor and percent change.
    c. Define a function that gives the number of coyotes in terms of the number of months that have elapsed.
    d. Re-write your function from part (c) so that it gives the number of coyotes in terms of the number of *years* that have elapsed.
    e. How long will it take for the number of coyotes to be one-half of the original number?

37. Sales of music CDs decreased by 6% each year over a period of 8 years.
    a. Determine the decay factor for 1 year.
    b. By what total percent did sales of the CDs change during the 8-year time period?
    c. How long will it take for the CD sales to be half of what they were at the beginning of this period? *If necessary, assume the trend continues in the future.*

38. Suppose the population of a town increased or decreased by the following percentages. For each situation, find the *annual percent change* of the population.
    a. Increases by 60% every 12 years
    b. Decreases by 35% every 7 years
    c. Doubles in size every 6 years
    d. Increases by 4.2% every 2 months
    e. Decreases by 7% every week

39. The number of asthma sufferers in the world was about 84 million in 1990 and 130 million in 2001. Let N represent the number of asthma sufferers (in millions) worldwide t years after 1990.
    a. Define a function that expresses N as a linear function of t. Describe the meaning of the slope and vertical intercept in the context of the problem.
    b. Define a function that expresses N as an exponential function of t. Describe the meaning of the growth factor and the vertical intercept in the context of the problem.
    c. The world's population grew by an annual percent change of 3.7% over those 11 years. Did the percent of the world's population who suffer from asthma increase or decrease?
    d. What is the long-term implication of choosing a linear vs. exponential model to make future predictions about the number of asthma sufferers in the world?

40. In the second half of the 20th century, the city of Phoenix, Arizona exploded in size. Between 1960 and 2000, the population of Phoenix increased by 2.76% each year. In the year 2000, the population of Phoenix was determined to be 1.32 million people.
    a. Define an exponential function f that gives the population P in Phoenix (in millions of people) where the input values t represent the number of years after the year 2000.
    b. Sketch a graph of this function.
    c. What input to your function will give the population of Phoenix in 1972? 1983? 1994? According to your function, approximately how many people lived in Phoenix in 1972? 1983? 1994?
    d. In what year did the population reach 1,000,000 people?
    e. The change in the population of Phoenix is increasing for equal changes in time. Illustrate this on the graph of f for at least 3 different equal intervals of time.
    f. What is the domain and range of this function in the context of the problem?
    g. Now, define an exponential function, g, modeling the population of Phoenix in millions of people n years after 1960. Sketch a graph of this function.
    h. How does the function you created in part (g) compare to the original function created in part (a)? How are the graphs of the two functions similar and different? Explain your reasoning.
    i. What is the domain and range of the function g, found in part (g)?

41. In each of the following equations, solve for the variable. That is, find the value of the variable that makes the equation true.
    a. $2(x-3)^{1/5}=-4$ b. $4(y-1)^{1/7}=3^{1/7}$ c. $(3x-2)^{7/3}=4$ d. $4(y+5)^{-2/3}=1$

V. *n*-UNIT GROWTH AND DECAY FACTORS ($n \neq 1$) (TEXT: S3)

42. For each function given below, find the initial value, specified factors and percent changes.
    a. $f(x)=2^{x/3}$
        i) Initial Value:
        ii) 1-unit Growth Factor:
        iii) 1-unit Percent Change:
        iv) 4-unit Growth Factor:
        v) 1/5-unit Growth Factor:

    b. $f(x)=5\cdot(0.98)^{x/4}$
        i) Initial Value:
        ii) 1-unit Decay Factor:
        iii) 1-unit Percent Change:
        iv) 3-unit Decay Factor:
        v) 1/2-unit Decay Factor:

    c. $f(x)=(0.25)^{2x}$
        i) Initial Value:
        ii) 1-unit Decay Factor:
        iii) 1-unit Percent Change:
        iv) 5-unit Decay Factor:
        v) 1/3-unit Decay Factor:

    d. $f(x)=3\cdot(1.6)^{5x}$
        i) Initial Value:
        ii) 1-unit Growth Factor:
        iii) 1-unit Percent Change:
        iv) 3-unit Growth Factor:
        v) 1/4-unit Growth Factor:

43. Determine the specified growth factor for the various unit intervals for the following exponential functions represented in the table below. Write your answers in exponential and decimal form (round to the nearest thousandth). In addition, determine the initial value and define the exponential function that models the data.

a.

x	1	3	5	7
$f(x)$	512	384	288	216

    i) Initial Value:
    ii) 1-unit Decay Factor:
    iii) 1-unit Percent Change:
    iv) 3-unit Decay Factor:
    v) 1/3-unit Decay Factor:
    vi) Function Formula:

b.

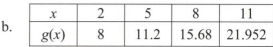

x	2	5	8	11
$g(x)$	8	11.2	15.68	21.952

    i) Initial Value:
    ii) 1-unit Growth Factor:
    iii) 1-unit Percent Change:
    iv) 2-unit Growth Factor:
    v) 1/4-unit Growth Factor:
    vi) Function Formula:

44. Determine the specified decay factor for the various time intervals for the following exponential function represented in the graph. Write your answers in exponential and decimal form (round to the nearest thousandth). Lastly, determine the function that models the data.

    a. 3-unit Decay Factor:
    b. 5-unit Decay Factor:
    c. 1-unit Decay Factor:
    d. 0.6-unit Decay Factor:
    e. Initial Value:
    f. Function Formula:

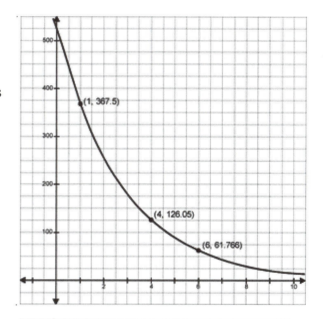

45. Determine the specified growth factor for the various time intervals for the following exponential function represented in the graph below. Write your answers in exponential and decimal form (round to the nearest thousandth). Lastly, determine the function that models the data.

    a. 2-unit Growth Factor:
    b. ½-unit Growth Factor:
    c. 1-unit Growth Factor:
    d. 2.5-unit Growth Factor:
    e. 5-unit Growth Factor:
    f. Initial Value:
    g. Function Formula:

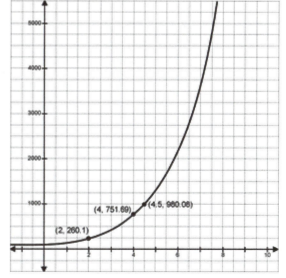

46. Determine the specified growth factor for the various time intervals for the following function. Write your answers in exponential and decimal form (round to the nearest thousandth).

The mass of bacteria in an experiment at time t days after its start is given by $f(t) = 95(4)^{t/5}$.

    a. 5-day growth factor: b. 1-day growth factor:

    c. Circle the statement(s) below that describes the behavior of the function above and give reasoning for why the statement(s) are true:

        i.   An initial mass of 95 μg of bacteria quadruples every 1/5-day.

        ii. An initial mass of 95 μg of bacteria quadruples every 5-days.

        iii. An initial mass of 95 μg of bacteria increases by 50% every 1-day.

47. Determine the specified decay factor for the various time intervals for the following function represented in the table below. Write your answers in exponential and decimal form (round to the nearest thousandth).

The number of buffalo in a wildlife preserve at time t months after its initial measure is given by $f(t) = 49(0.97)^{2t}$.

    a. 1-month decay factor: b. ½-month decay factor:

    c. Circle the statement(s) below that describe(s) the behavior of the function above:

        i.   An initial number of 49 buffalo decreases by 3% every ½-day.

        ii. An initial number of 49 buffalo decreases by 3% every 2-days.

        iii. An initial number of 49 buffalo decreases by 5.91% every 1-month.

48. Determine the specified growth or decay factor for the various time intervals for the following function described in the context below. Write your answer in exponential and decimal form (round to the nearest thousandth).

    a. The population of Jackson has 57,421 people and is growing by 4.78% every 4 years.

        i)   8-year growth factor:

        ii) What is the percent change (increase) over 8-year period?

        iii) 1-year growth factor:

        iv) What is the percent change (increase) over 1-year period?

    b. The amount of drug B in your body decreases by 18% every 3 hours.

        i)   24-hour decay factor:

        ii) What is the percent change (decrease) over a 24-hour period?

        iii) 1-hour decay factor:

        iv) What is the percent change (decrease) over a 1-hour period?

49. Filling in the Missing Pieces: Use the data to fill in the blanks.

    a. Determine whether the following tables represent a linear or exponential function.

    b. Fill in the blanks of the tables.

    c. Define a function for each table.

Table i	
Input	Output
−2.0	−10.0
−1.5	
−1.0	−7.5
−0.5	
0.0	−5.0
0.5	
1.0	−2.5
1.5	
2.0	0

Table ii	
Input	Output
−4.0	0.1111
−3.0	
−2.0	0.3333
−1.0	
0.0	1.0
1.0	
2.0	3.0
3.0	
4.0	9.0

Table iii	
Input	Output
0.0	132.0
0.5	
1.0	29.04
1.5	
2.0	6.3888
2.5	
3.0	1.4055
3.5	
4.0	0.309
4.5	

50. Rewrite each expression in expanded form.
    a. $(x+3)^2$
    b. $(6x-7)^2$
    c. $(2x+1)^3$
    d. $(x-2)^2(x+4)^2$
    e. $\left(2x^4\right)^3(-3x+5)^2$
    f. $(4x)^2(-2x)^3(x+1)^2$

VI. COMPOUNDING PERIODS & COMPOUND INTEREST FORMULA (TEXT: S4)

51. The table below illustrates the value of an investment from the end of the 7th compounding period to the end of the 10th compounding period. Use this table to answer parts (a) through (f) below.

Number of compounding periods p	Investment value
7	$7,105.57
8	$7,209.31
9	$7,314.57
10	$7,421.36

    a. Verify that the data in the table represents exponential growth.
    b. How does the investment value change from the end of the 7th to the end of the 9th compounding period? From the end of the 8th to the end of the 10th compounding period?
    c. What is the value of the investment after 3 compounding periods? After 20 compounding periods?
    d. Define a function f that expresses the investment as a function of the number of compounding periods p. Note: p is defined to be the values $\{0,1,2,3,...\}$. Explain the meaning of each of the values in your function.
    e. Construct a graph of the function defined in part (d). Label the axes appropriately, then pick one coordinate point from the graph and explain what it represents.
    f. Determine the number of compounding periods until the investment reaches $400,000.

52. For each of the following accounts, determine the percent change per compounding period. Give your answer in both decimal and percentage form.
    a. 5% APR compounded monthly
    b. 6.7% APR compounded quarterly
    c. 3.2% APR compounded daily
    d. 8% APR compounded each hour

53. For each of the accounts in the previous question, give the growth factor per compounding period, then give the annual growth factor and the annual percent change (APY).

54. If you invest \$1,200 in a CD with an APR of 3.5% compounded monthly, the following expression will calculate the value of the CD in 6 years: $1200\left(1+\dfrac{0.035}{12}\right)^{12(6)}$

    Explain what each part of the expression represents in this calculation.

    a. $\dfrac{0.035}{12}$

    b. $1+\dfrac{0.035}{12}$

    c. $12(6)$

    d. $\left(1+\dfrac{0.035}{12}\right)^{12}$

    e. $\left(1+\dfrac{0.035}{12}\right)^{12(6)}$

55. Write an expression that would calculate the value of the following account after 14 years: \$8100 is invested at an APR of 4.8% compounded semiannually (twice per year).

56. Write an expression that would calculate the value of the following account after 30 years: \$16,000 is invested at an APR of 3.5% compounded daily.

57. An investment of \$10,000 with Barnes Bank earns a 2.42% APR compounded *monthly*.
    a. Define a function that gives the investment's value as a function of the number of years since it began.
    b. Determine the investment's value after 20 years.
    c. Determine the annual growth factor, and annual percent change (APY).
    d. Determine how long will it take the investment to double.
    e. Another bank says they will pay you the same interest (2.42% APR), but compounded daily. What would be the value of a \$10,000 investment with this bank, leaving it there for 20 years? Compare this answer to your account with Barnes Bank.

58. You just won \$1000 in the lottery and you decide to invest this money for 10 years.
    a. Which of the three different accounts would you choose to invest your \$1000? Provide calculations for each account and justify your reasoning.
       - Account #1 pays 14% interest each year.
       - Account #2 pays 13.5% interest per year, compounded monthly.
       - Account #3 pays 13% interest per year, compounded weekly.
    b. Describe how your money increases as time increases for each account. Be sure to incorporate the annual growth factor and annual percent change into your description!

59. Karen received an inheritance from her grandparents and wants to invest the money. She is offered the following options of accounts to invest into:
       - 4.5% APR, compounded semi-annually
       - 4.3% APR, compounded daily
    a. Which option should she choose? Explain your reasoning using the accounts' APY.
    b. If Karen decides on the first option, by what *percent* will her investment have *increased* after 10 years?

60. John decides to start saving money for a new car. He knows he can invest money into an account which will earn a 6.5% APR, compounded weekly, and would like to have saved \$10,000 after 5 years.
    a. How much money will he need to invest into the account now so that he has \$10,000 after 5 years?
    b. Determine the APY (Annual Percent Yield) for the account.
    c. Determine the 5-year percent change for the account.

VII. INVESTMENT ACTIVITY: FOCUS ON FORMULAS AND MOTIVATING e (TEXT: S4)

61. Suppose $4500 is initially invested into an account with an APR of 7%.
    a. Determine the value of the account at the end of the given years and the APY for each type of compounding per year.

Year	Compounded Monthly	Compounded Daily	Compounded Continuously
1			
2			
4			
10			
30			
Determine the APY for each situation			

    b. Compare the short-term vs. the long-term impact of increasing the number of times interest is compounded each year.

62. Determine the value of each of the following accounts after 14 years.
    a. Initial investment is $4000 with a 7.1% APR compounded continuously
    b. Initial investment is $750 with a 3.3% APR compounded continuously
    c. Initial investment is $2000 with a 5% APY compounded continuously

63. Suppose that $2000 is initially invested in an account with an APR of 3.4% compounded continuously.
    a. Determine the value of the account at the end of 5 years.
    b. Write a function that models the value of the account at the end of t years.
    c. What is the annual percent change (APY) of the account? What does this value represent?
    d. What is the percent change over 10 years for this account?

64. Suppose $18,000 is initially invested in an account with an APR of 5.1%.
    a. Does it make a bigger impact going from compounding interest annually to monthly or going from compounding interest daily to continuously?
    b. What is the annual percent change (APY) for each of the four compounding methods listed in part (a)?
    c. If interest is compounded continuously, how much interest does the account earn over the first 10 years?

65. Suppose an initial population of 32 million people increases at a continuous percent rate of 1.9% per year since the year 2000.
    a. Determine the function that gives the population in terms of the number of years since 2000.
    b. Determine the population in the year 2019.
    c. What is the annual growth factor for this context? Explain two ways we can determine this value.
    d. What does your answer to part (c) represent in this context?

66. Carbon-14 is used to estimate the age of organic compounds. Over time, carbon-14 decays at a continuous percent rate of 11.4% per thousand years from the moment the organism containing it dies. Carbon-14 is typically measured in micrograms.
    a. What quantities are changing in this situation? What quantities are not changing?
    b. Define a function that gives the amount, at any moment, of Carbon-14 remaining in a piece of wood that starts out with 150 micrograms of Carbon-14. (Remember that Carbon-14's decay rate is per thousand years.)
    c. Construct a graph of this function. Be sure to label your axes!
    d. What is the percent change every 1000 years?
    e. What is the percent change every 5000 years? Explain your reasoning.

VIII.   THE INVERSE OF AN EXPONENTIAL FUNCTION (TEXT: S5)

Log Property #1: $\log_b (m \cdot n) = \log_b m + \log_b n$ for any $b > 0$, $b \neq 1$ and $m, n > 0$

Log Property #2: $\log_b \left(\dfrac{m}{n} \right) = \log_b m - \log_b n$ for any $b > 0$, $b \neq 1$ and $m, n > 0$

Log Property #3: $\log_b (m^c) = c \cdot \log_b m$ for any $b > 0$, $b \neq 1$ and $m, n > 0$

67. Estimate the following logs. Explain why your estimate makes sense:
    a. $\log_4 (60)$ b. $\log_3 (143)$ c. $\log_8 \left(\frac{1}{64} \right)$ d. $\log_9 (27)$

68. Find the unknown in each of the following equations. Estimate if necessary. For parts g and h, consider z to be greater than 0.
    a. $\log_3 (20) = y$ b. $2(\log_5 (625)) = y$ c. $\log_5 (1) = y$ d. $\log_2 (30) = y$
    e. $\log_2 (x) = 4.2$ f. $\log_7 (x) = -2$ g. $\log_z (343) = 3$ h. $\log_z (1) = 0$

69. For each of the following:
    • Write the expression as a single logarithm.
    • Evaluate to a single number or estimate the value of the expression.
    a. $\log_5 (6.25) + \log_5 (100) + \log_5 (25)$ b. $\log_4 \left(\frac{1}{32} \right) + \log_4 \left(\frac{1}{8} \right)$ c. $2 \cdot \log_6 (12) + \log_6 (4)$
    d. $\ln(6) + \ln(3) - \ln(2)$ e. $\log_5 (6) - \log_5 (100)$ f. $\frac{1}{2} \left(\log_5 (8) + \log_5 (3) \right)$

70. Solve each of the following for x. (Hint: Use the properties discussed in class or your understanding of logarithms.)
    a. $\log(0.01x) = 0$ b. $\log_5 (25x^2) = 6$ c. $\ln \left(\frac{x}{10} \right) = 4$

71. Solve each of the following for x.
    a. $\log_2 (2 + x) + \log_2 (7) = 3$ b. $\ln(3x^2) - \ln(5x) = \ln(x + 9)$

72. Rewrite each of the following as sums and differences of a logarithm of some number.
    a. $\log_7 \left(\dfrac{4}{y} \right)$ b. $\log_2 (x^4 \cdot 12)$ c. $\log_5 \left(\dfrac{10x^3}{y^5} \right)$

73. Rewrite each of the following exponential equations in logarithmic form.
    a. $y = 11^x$
    b. $y = 1.7(3.2)^t$
    c. $y = 200(1.0027)^{12t}$

74. Graph each of the following functions.
    a. $f(x) = \log_3(x)$
    b. $g(x) = \log_5(x)$

75. a. Sketch the following two functions on the same set of axes: $f(x) = \log(x)$ & $g(x) = \log_5(x)$
    b. Consider the rates of change of each of the functions. Explain why $g(x)$ has a greater rate of change than $f(x)$ for all values of $x > 1$.
    c. For what values of x is $\log(x) > \log_5(x)$? Explain your reasoning.

IX. SOLVING EXPONENTIAL AND LOGARITHMIC EQUATIONS (TEXT: S5)

76. The amount of an investment is determined by the function f defined by $f(t) = 4186.58(1.025)^t$. Algebraically, determine when (in compounding periods, t) the investment will reach $1,000,000.

77. An initial amount of 120 mg of caffeine is metabolized in the body and decreases at a continuous percent rate of 21% per hour.
    a. Define an exponential function that gives the amount of caffeine remaining in the body after t hours.
    b. How many hours will it take for the amount of caffeine to reach half of the initial amount? (Solve this both graphically and symbolically to verify your answers.)

78. An initial investment of $6000 is made to an account with an APR of 4.7%.
    a. If interest is compounded monthly, how many years will it take for the account balance to be $10,290.32? Solve algebraically and check your answer.
    b. If interest is compounded continuously, how many years will it take for the account balance to be $9,510.15?

79. The amount of medicine in a patient's bloodstream for reducing high blood pressure decreases at a continuous percent rate of 27% per hour. This medicine is effective until the amount in the bloodstream drops below 1.2 mg. A doctor prescribes a dose of 85 mg.
    a. Define the function A that models the amount of medicine remaining in the bloodstream after t hours.
    b. About how long until 45 mg of medicine remains in the patient's bloodstream? 1.2 mg?

80. The rate at which a wound heals can be modeled by the exponential function $f(n) = Ie^{-0.1316n}$ where I represents the initial size of the wound in square millimeters and $f(n)$ represents the size of the wound after n days. This function assumes no infection is present and no antibiotic ointment is used to speed healing.
    a. Suppose you scrape your knee and get a wound 300 square millimeters in size. Define a function to model the size of the wound with respect to time.
    b. How large will the wound be after one week?
    c. How long will it take for the wound to be 20% of its original size?
    d. You want to know how long it will take to reduce the size of the wound to 20% of the size you determined in part (c). How will the amount of time it takes to do this compare to the amount of time it took to reduce the wound to 20% of its original size? Explain your reasoning.

81. Given the function $f(t) = 10(0.71)^t$, where t is in years:
    a. What is the annual percent change?
    b. Convert the function $f(t) = 10(0.71)^t$ into the equivalent form $f(t) = ae^{kt}$
    c. What is the continuous annual percent rate?

82. The town of Gilbertville increased from a population of 3,562 people in 1970 to a population of 9,765 in 2000.
    a. Define an exponential function that models the town's population as a function of the number of years since 1970.
    b. What is the annual percent change?
    c. Use your function to predict the town's population in 2019.
    d. According to your function, when will the town's population reach 40,000 people? (Answer this question both graphically and symbolically.)
        e. After how many years from *any* reference year will the population triple?

*1. Examine the bottle given below and imagine the bottle filling with water. Make a possible rough sketch of a graph of the height of the water in the bottle as a function of the volume of the water in the bottle. (Hint: it may be helpful to imagine adding equal amounts of volume of water and consider how the height of the water in the bottle will change.)

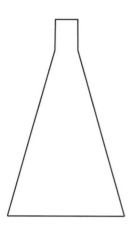
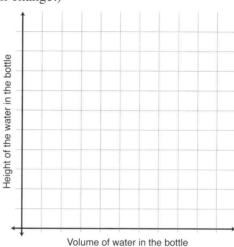

*2 a. Using the bottle animation, imagine filling the bottle with water and complete the following table. Create a revised sketch of the graph of the height of the water in the bottle as a function of the volume of the water in the bottle.

Volume of Water	Height of Water

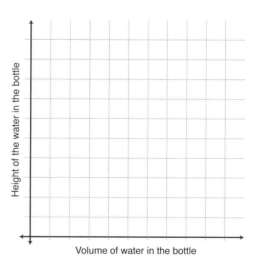

b. What does your graph convey about how the height of the water in the bottle and the volume of water in the bottle change together? (Be sure to explain how the amounts of increase in the volume and amounts of increase in the volume change together both before and after the neck of the bottle.)

c. Calculate the following average rates of change of the height of the water with respect to the volume of the water and explain what these values mean in the context of the problem.

   i. The average rate of change as the volume of the water increases from 1 to 2.

   ii. The average rate of change as the volume of the water increases from 2 to 3.

   iii. The average rate of change as the volume of the water increases from 3 to 4.

d. What do you notice about the average rates of change for these successive subintervals? Explain why your answer makes sense relative to the shape of the bottle.

3. a. Use the graph of the function *f* that represents the *height of water in the bottle* in terms of the *volume of water in the bottle* to complete the following table.

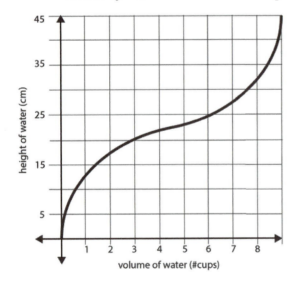

Volume of Water	Height of Water	Average Rate of Change of height with respect to volume
0		
1		
2		
3		
4		
5		
6		
7		
8		
9		

b. Use the graph of *f* and table of values for *f* to answer the following questions.

  i. On what interval(s) of the function *f* is the average rate of change decreasing? Describe the thinking you used to arrive at your answer.

  ii. What predictions can you make about the shape of the bottle on this interval(s)?

  iii. Describe how *changes in the average rate of change of f on an interval* relate to the shape of the graph of *f* on intervals of the function's domain. (Hint: When the average rate of change of the height with respect to volume is increasing/decreasing on an interval…)

  iv. Draw a picture of a bottle that would produce the volume-height relationship conveyed in the table and graph of *f*. Label landmarks on your bottle (and the graph in part (a)) where the *rate of change of the height with respect to volume* changes from either increasing to decreasing, or decreasing to increasing.

*1. At 5:00 pm Karen started walking from the grocery store back to her house.
   a. Fill in the below table by determining the number of feet Karen is from home, d. Then use the information in the table to determine the average rate of change of the number of feet Karen is from home with respect to time (measured in minutes since Karen started walking) on the specified intervals.

Change in the number of minutes since Karen started walking = Δt	Number of minutes since Karen started walking = t	Number of feet Karen is from home = d	Change in the number of feet Karen is from home = Δd	Average rate of change of Karen's distance with respect to time
	0	118		
			-1.5	
	0.5			
			-3.2	
	1			
			-6.5	
	1.5			
			-7.1	
	2			

   b. Sketch a graph of the number of feet Karen is from home in terms of the number of minutes since Karen started walking. *(Be sure to label your axes)*

   c. Is the graph you drew above concave up, concave down, or some combination of both on the interval $0 < t < 2$? Use the table above to justify your answer.

   d. Describe how the quantities *number of minutes since Karen started walking* and the *number of feet Karen is from home* change together.

*2. You have finished your bath and pull the plug from the drain. As the water is draining you record the number of minutes *n* and the height of the water *h* (in centimeters) of the water in the tub since the plug was pulled.

   a. Fill in the below table by determining the change in the number of minutes Δn and the change in the height of water Δh over the specified intervals. Then use this information to determine the average rate of change of the height with respect to time (measured in minutes since the plug was pulled) on the specified intervals.

Change in the number of minutes elapsed $= \Delta n$	Number of minutes elapsed = n	Height of water in centimeters $= h$	Change in the height of water in centimeters $= \Delta h$	Average rate of change of height with respect to #minutes elapsed
	1	50		
	1.5	45.75		
	2.6	37.83		
	4.5	29.83		
	7	20.83		

   b. Sketch a graph of the height of the water in centimeters with respect to the number of minutes elapsed since you pulled the plug from the drain. *(Be sure to label your axes)*

   c. Is the graph you drew above concave up, concave down, or some combination of both on the interval from $1 < n < 7$? Use the table above to justify your answer.

   d. Use the table you created in part (a) to explain how patterns in the *average rate of change of height with respect to number of minutes* relate to the concavity of the graph. Describe why it was valuable to examine the *average rate of change of the height* with respect to the *number of minutes* elapsed on intervals of the domain.

   e. Describe how the quantities *number of minutes elapsed since you pulled the plug from the drain* and the *height of the water in centimeters in the tub* change together as the water drains from the tub.

*1. The function g represents the height of a ball above the ground (measured in feet) in terms of the amount of time t (measured in seconds) since the ball was thrown upward from the roof of a building.

    a.   Complete the table below by determining the average rate of change of the height of the ball above the ground (measured in feet), with respect to the amount of time t (in seconds) since the ball was thrown.

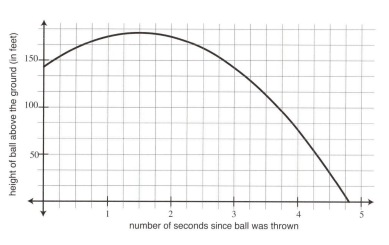

t	$g(t)$	Average Rate of Change
0	145	
0.5	165	
1	177	
1.5	181	
2	177	
2.5	165	
3	145	

    b.   What pattern do you notice in how the *average rates of change of the height of the ball with respect to number of seconds since ball was thrown* changes, when considering 0.5 second increments of time?

    c.   What does the point (2.5, 165) represent in the context of this situation?

*2. A second ball is thrown from a platform that is exactly 5 feet above the point where the ball modeled by function g is thrown. Let h be the function that models the distance-time relationship of this second ball. (Assume the second ball follows the same trajectory as the first ball.)

    a.   Complete the table at right.

    b.   What does it mean for the two balls to follow the same trajectory?

    c.   How do the output values of function h compare with the output values of function g on the domain of h? Sketch function h on the same axes that g is sketched on above.

t	$h(t)$	Average Rate of Change
0		
0.5		
1		
1.5		
2		
2.5		
3		

d. How do the average rates of change of functions h and g compare on successive ½ second intervals of the domains of h?

e. Express the function h in terms of the function g.

*3. Let f be the function that describes the distance-time relationship of a ball whose distance above the ground is always 3 times as large as the initial ball's distance above the ground (modeled by function g).

a. Complete the table at right.

b. How do the output values of the function f compare with those of function g?

c. How do the average rates of change of functions f and g compare on successive ½ second intervals of the domain of f?

t	$f(t)$	Average Rate of Change
0		
0.5		
1		
1.5		
2		
2.5		
3		

d. Express the function f in terms of the function g.

*4. Let k be the function that models the distance-time relationship of a ball whose distance above the ground is always 3 times as large as the initial ball's distance above the ground (modeled by function g). This ball was also thrown from a platform 5 feet higher than the ball modeled by function g.

a. Complete the table at right.

b. How do the output values of the function k compare with those of function g?

c. How do the average rates of change of functions k and g compare on successive ½ second intervals of the domain of k?

t	$k(t)$	Average Rate of Change
0		
0.5		
1		
1.5		
2		
2.5		
3		

d. Express the function k in terms of g.

5. Let p be the function that is the reflection of the function g across the horizontal-axis.

   a. Complete the table at right.

   b. Does the function p make sense as a model of the time-distance relationship of a ball that is thrown from the ground? Explain.

   c. How do the output values of the function p compare with those of function g?

t	$p(t)$	*Average Rate of Change*
0		
0.5		
1		
1.5		
2		
2.5		
3		

   d. How do the average rates of change of functions p and g compare on successive ½ second intervals of the domain of p?

   e. Express the function p in terms of g.

*6. Explain what the function w defined by $w(x) = 2g(x) + 7$ represents in the context of the distance-time relationship of the ball.

*1. Use your graphing calculator to sketch a graph of the following quadratic functions and compare the behavior and properties of these functions. For each of these functions determine the roots (*x*-intercepts) of the function.

a. $f(x) = x^2$

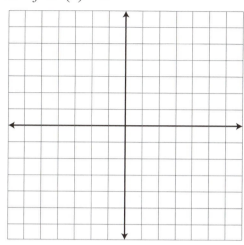

b. $g(x) = 3x^2$

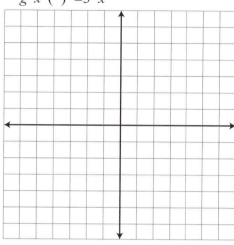

c. $h(x) = 3x^2 - 27$

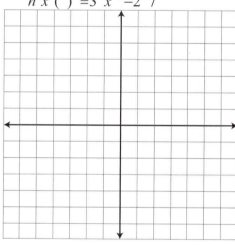

d. $k(x) = 3(x - -7)^2$

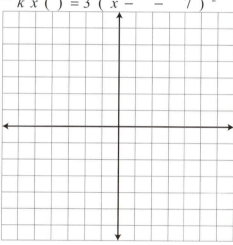

e. $p(x) = (2x - 9)(x + 4)$

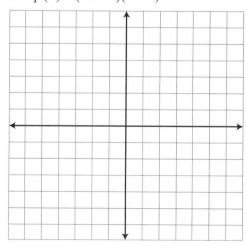

*2. Use your graphs from (1) to answer the following questions.

    a. How do you determine the number of roots (*x*-intercepts) of a quadratic function?

    b. Is it possible for a quadratic function to be both concave up and concave down? Explain.

    c. Is there always a maximum or minimum value in a quadratic function? Explain.

    d. How do you determine if a quadratic function will be concave up or concave down?

    e. What is the relationship between the *x*-coordinate of the maximum/minimum value of a quadratic function *g* and the roots of *g*?

*3. Let $f(x) = (x+7)(x-3)$

    a. Use algebraic methods to determine the roots of *f*.

    b. Determine the *x*-coordinate of the vertex.

Given a quadratic function in standard form, $f(x) = ax^2 + bx + c$, the **quadratic formula** determines the roots of a quadratic function and is defined by;

$$x = \frac{-b}{2a} \pm \frac{\sqrt{b^2 - 4ac}}{2a}$$

*4a. Explain why the quadratic formula makes sense. (You do not need to figure out how to derive the second term, but think about what this second term represents).

b. Assume that for a given quadratic function, f, the solution to the quadratic formula is $x = -4 \pm 3.25$. Create a possible sketch of this quadratic function and illustrate the vertex and roots of this quadratic function.

5. An ice cream shop finds that its weekly profit P (measured in dollars) as a function of the price x (measured in dollars) it charges per ice cream cone is given by the function
$$P = k(x) = -125x^2 + 670x - 125$$

a. Determine the maximum weekly profit and the price of an ice cream cone that produces that maximum profit.

b. If the cost of the ice cream cone is too low then the ice cream shop will not make a profit. Determine what the ice cream shop needs to charge in order to break even (make a profit of $0.00).

c. If the cost of the ice cream cone is too high then not enough people will want to buy ice cream and so the weekly profit will be $0.00. Determine what the ice cream shop would have to charge in order for this to happen.

d. Describe the meaning of $g(x) = k(x-2)$. Does this function have the same maximum profit as k? If not, what is the price per ice cream cone that produces the maximum profit of the function k? Explain.

e. Describe the meaning of $h(x) = k(x) - 2$. Does this function have the same maximum profit as k? If not, what is the price per ice cream cone that produces the maximum profit of the function k? Explain.

*1. Let $f(x) = (x+1)(x-3)^2$.

    a.   Use algebraic methods to find the roots (*x*-intercepts) of *f*.

    b.   What do the roots of a polynomial function represent?

    c.   Explain why the zeros of a polynomial function occur where each factor is equal to zero.

    d.   WITHOUT USING A CALCULATOR determine the interval(s) on which the output of the function *f* is:
       i.   positive (Show work)

       ii.   negative (Show work)

    e. Construct a rough sketch of the graph of *f*.

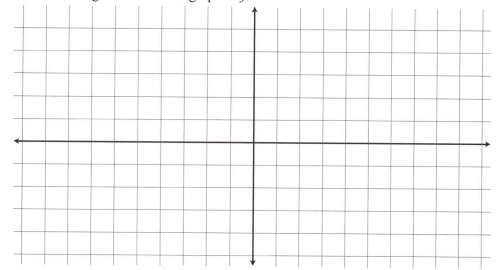

3. How do the functions f and h compare, given that $f(x) = (x+1)(x-3)^2$ and $h(x) = 2(x+1)(x-3)^2$?

*4. How do the functions h and k compare, given that $h(x) = 2(x+1)(x-3)^2$ and $k(x) = -2(x+1)(x-3)^2$?

*5. Let $f(x) = x^3$, $g(x) = x^5$, $h(x) = x^2$, $j(x) = x^8$. (Note that a function of the form $k(x) = ax^n$, given that a is a real numbers and n is a positive integer is called a power function.)
   a. Describe how each power function varies as:
      i. x increases without bound (also written $x \to \infty$) Explain.

      $x \to \infty$, $f(x) \to$ _____

      $x \to \infty$, $g(x) \to$ _____

      $x \to \infty$, $h(x) \to$ _____

      $x \to \infty$, $j(x) \to$ _____

      ii. x decreases without bound (also written $x \to -\infty$) Explain.
      $x \to -\infty$, $f(x) \to$ _____

      $x \to -\infty$, $g(x) \to$ _____

      $x \to -\infty$, $h(x) \to$ _____

      $x \to -\infty$, $j(x) \to$ _____

   b. What general statements can you make about how the exponent on a power function impacts the behavior of the function?

*6. Compare $f(x) = x^3 + 4x^2 - 6x - 12$ and $g(x) = x^3$.

7. For each polynomial function, identify the leading term and describe the function's end behavior. (Remember to consider the role of the leading coefficient when determining a function's end behavior).

   a. $f(x) = 9x^2 - 4x - 2x^4$

   b. $g(x) = -7(x-5)(x^2 + 1)$

*8. Answer the following questions given the graph of the function g below.

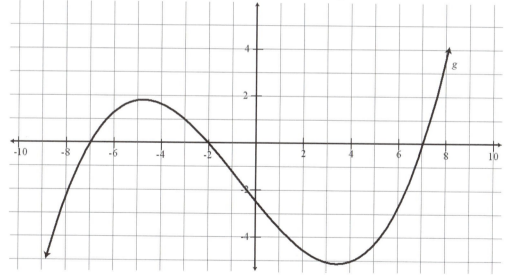

   a. What are the roots of g?

   b. Evaluate $g(0)$

   c. On what interval(s) is the function g concave up?

   d. On what interval(s) is the function g concave down?

e. $\lim\limits_{x \to \infty} g(x)$

   (What does *g(*x) approach, as *x* grows without bound?)

f. $\lim\limits_{x \to -\infty} g(x)$

   (What does *g(*x) approach, as *x* decreases without bound?)

g. Does the function have odd or even degree? Explain.

h. Are there any factors with an even power? Explain.

I. THE BOTTLE PROBLEM (TEXT: S2, 3, 4, 5)

1. Consider a spherical fish bowl that is being filled with water.

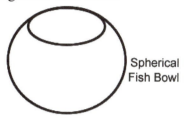

Spherical
Fish Bowl

    a. Describe how the height of the water and volume of water in the bowl vary together for equal amounts of volume of water added to the bottle.

    b. Construct a graph that represents the height of the water in the bowl as a function of the volume of the water in the bowl.

    c. Describe how the rate of change of the height of the water in the bowl with respect to the volume of water in the bowl changes as the volume of the water in the bowl varies from 0 to the volume of water that corresponds to a full bowl.

2. For each of the scenarios below, construct a graph of the height of the water in the bottle as a function of the volume of water in the bottle. Then, make an illustration of a bottle that would produce a height-volume graph with that general shape.

    a. Scenario #1: A bottle in which the height of the water in the bottle increases at a constant rate with respect to the volume of water in the bottle.

    b. Scenario #2: A bottle in which the height of the water in the bottle increases at a constant rate with respect to the volume of water in the bottle, then increases at a decreasing rate with respect to the volume of water in the bottle.

    c. Scenario #3: A bottle in which the height of the water in the bottle increases at an increasing rate with respect to the volume of water in the bottle, then increases at a constant rate with respect to the volume of water in the bottle, then increases at a decreasing rate with respect to the volume of water in the bottle.

3. Joe is pouring water in a pitcher. He pours 3 cups of water into the pitcher and notices the height of the water in the pitcher is increasing at a decreasing rate with respect to volume. He then pours 2 more cups of water into the pitcher and notices the height of the water in the pitcher is increasing at a constant rate with respect to the volume of water. Joe adds 5 more cups of water and sees that the height of the water in the pitcher is increasing at an increasing rate with respect to volume.

    a. As the first 3 cups of water are poured into the pitcher describe how the height of the water and the volume of water in the bottle vary together.

    b. Construct a graph of the height of the water in the pitcher as a function of the volume of water in the bottle. Be sure to label your axis.

    c. Make a careful sketch of the pitcher that would produce a height-volume graph with the general shape determined in part (b).

4. Below are four graphs that represent the height of water in a bottle as a function of the volume of water in the bottle. For each graph:
   i.   Describe how the height of the water and volume of water in the bottle vary together for equal amounts of volume of water added to the bottle.
   ii. Describe how the volume of water and rate of change of height of water in the bottle vary together.
   iii. Construct a careful sketch of the bottle based on the graph and your descriptions from part (a) and (b). Include landmarks on both the bottle and graph to show the changes of the graph relative to the shape of the bottle.

a.

b.

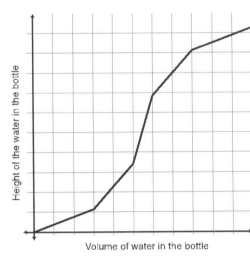

c.

d.

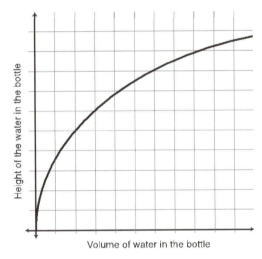

5. Given that *h* represents the height of the water in a bottle as a function of the volume of water in the bottle, the graph of *h* will always be increasing (rising as you move from left to right on the graph) since increasing the volume of water in the bottle results in increasing the height of water in the bottle. The definition of an **increasing function** *f* is: A function *f* is said to be increasing if
$f(x_1) < f(x_2)$ whenever $x_1 < x_2$.
   a. Given that *g* is a function that defines the height of water in a bottle as a function of the volume of the water, *v*, explain the meaning of the statement $g(v_1) < g(v_2)$ whenever $v_1 < v_2$.
   b. Use mathematical symbols to convey that a function *w* is decreasing for all values of *x*. How does the graph of *w* change as *x* increases?

6. When graphing the height of the water in a bottle with respect to the volume of water in the bottle, **inflection points** on the height–volume graph correspond to where the bottle changes from getting narrower to getting wider (or vice-versa)

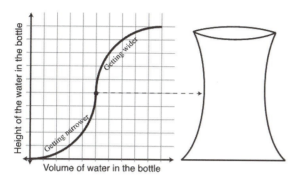 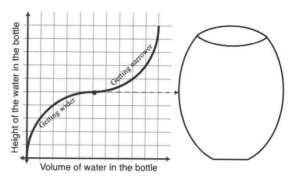

   a. Explain what the inflection point on the height-volume graph for the bottle on the left conveys about the changes in the volume of water in the bottle and the height of the water in the bottle as the bottle is being filled with water.
   b. Explain what the inflection point on the height volume graph for the bottle on the right conveys about the changes in the volume of water in the bottle and the height of the water in the bottle as the bottle is being filled with water.
   c. Explain how the volume of water in the bottle and the rate of change of the height of the water in the bottle with respect to the volume of water in the bottle change at the inflection points for the height-volume graphs above.

7. As a runner is moving around a quarter mile track a radar gun detects the direct distance of the runner from the starting line.

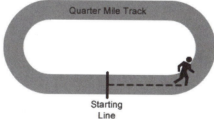

   a. Construct a graph that represents the direct distance (in yards) of the runner from the starting line in terms of the total distance (in yards) the runner has traveled around the track.
   b. At approximately what point(s) around the quarter mile track is the direct distance of the runner from the starting line at its maximum value?
   c. Provide a written description of how the direct distance (in yards) of the runner from the starting line varies with the distance (in yards) the runner has traveled around the track.
   d. Where is the runner on the track when her direct distance from the starting line (with respect to the distance she has traveled around the track) changes from increasing at a constant rate to increasing at a decreasing rate?
   e. Are there inflection points on your graph? If so, indicate approximately where they are on your graph and explain what they represent.

8. A skateboarder skates on a half-pipe like the one shown below.

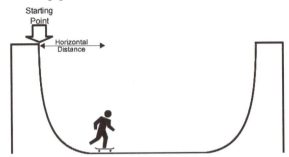

   a. Construct a graph that represents the skateboarder's horizontal distance from the starting point in terms of the total distance traveled by the skateboarder since leaving the starting point as the skateboarder goes to the far end of the half-pipe and back to the starting point.
   b. Where is the skateboarder on the half-pipe when his horizontal distance from the starting point is increasing at the greatest rate with respect to the total distance traveled from the starting point?
   c. Where is the skateboarder on the half-pipe when his horizontal distance from the starting point is decreasing at the greatest rate with respect to the total distance traveled from the starting point?
   d. Provide a written description of how the horizontal distance (in feet) from the starting point varies with the total distance traveled (in feet) since leaving the starting point.
   e. Are there any inflection points on your graph? If so, indicate where they are on the graph and explain what they represent.

9. For the following graphs determine whether the rate of change of the height of the water with respect to volume of the water is increasing, decreasing, or constant for successive intervals of the domain. Justify your reasoning.

   a.

   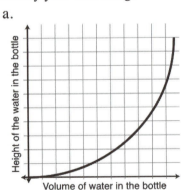

   b.

   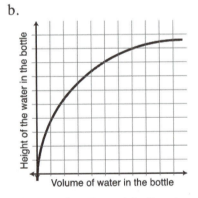

   c.

   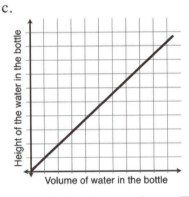

10. Watch the video titled "Exploring Co-Varying Quantities" and answer the following questions. (The video can be found in the online textbook, Module 5 p. 7)
    a. What conditions are necessary for there to be a linear relationship between the height of the water and the volume of water in the bottle?
    b. For a cylindrical bottle with shoulders (like the one considered in the video) how does the height of the water in the bottle co-vary with the volume of water in the bottle?
    c. What do sharp corners on a height-volume graph represent?
    d. What is a technique that can be used when analyzing the co-variation of two quantities?

11. Expand the following expressions as much as possible.
    a. $(x-3)(x+4)$ b. $-(x-7)^2$ c. $(x-4)^2(x-1)$
    d. $3(x-2)^2$ e. $-(3x-2)(x+4)(2x-1)$

II. CO-VARIATION AND CHANGING RATE OF CHANGE (TEXT: S4, 5)

12. Answer the questions below by using the graph of a function f that represents the amount of money in a banking account (measured in thousands of dollars) as a function of the number of years since 1995.

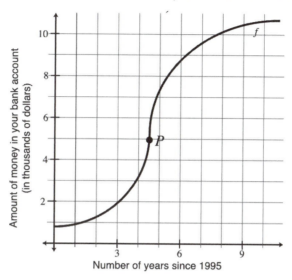

a. How did the amount of money in the bank account change from 0 to 5 years after 1995?

b. What does the point P on the graph convey about how the amount of money in the account is changing with time? (Hint: Examine how the amount of money in the account is changing for small time intervals prior to and after $t = 4.5$ years past 1995?)

c. The graph is concave down on the interval $4.5 < t < 10$. How is the amount of money in the account changing on this time interval? How is the rate of change of the amount of money in the account changing on this time interval?

13. Suppose $d = f(t) = 1.5^t$ represents the distance d (measured in inches) of a car from a stop sign in terms of the number of seconds t since the car started to move away from the stop sign.

a. Sketch a graph of the number of inches the car is from the stop sign in terms of the number of seconds since the car started to move away from the stop since. (Be sure to label your axes).

b. Determine the average rate of change of the distance of the car from the stop sign on the following time intervals.

    i. $0 \le t \le 2$

    ii. $2 \le t \le 4$

    iii. $4 \le t \le 6$

c. Describe what these average rates of change tell you about how the distance of the car from the stop sign is changing over the time interval from $t = 0$ to $t = 6$ seconds.

d. Is your graph concave up, concave down, or some combination of both on the interval $0 < t < 6$? Justify your answer from part (b) and (c).

14. The values in the table below represent the distance (measured in yards) of a dog from a park entrance as a function of the number of seconds since the dog entered the park.

Change in the number of seconds since the dog entered the park	The number of seconds since the dog entered the park	The distance (in yards) of a dog from the park entrance	Change in the distance (in yards) of a dog from the park entrance	Average rate of change of the dog's distance with respect to time
	0	0		
	3	26.25		
	3.2	30.976		
	4.5	77.344		
	7	271.25		

   a. Complete the table of values above.
   b. Sketch a graph of the number of yards the dog is from the park entrance in terms of the number of seconds since the dog entered the park. (Be sure to label your axes).
   c. Describe what these average rates of change tell you about how the distance of the dog from the park entrance is changing over the time interval from $t = 0$ to $t = 7$ seconds.
   d. Is your graph concave up, concave down, or some combination of both on the interval $0 < t < 7$? Justify your answer from part (a) and (c).

15. The values in the table below represent the distance (measured in yards) of a car from a stop sign as a function of the number of seconds since the car started to move away from the stop sign.

Number of seconds since the car started to move away from the stop sign	Distance (in yards) of a car from a stop sign
0	0
1	2
2	6
3	12
4	20

   a. Using the table of values above:
      i. Sketch a graph of the number of yards the car is from the stop sign in terms of the number of seconds since the car started to move away from the stop since. (Be sure to label your axes).
      ii. Represent the changes in the distance of the car from the stop sign from 0 seconds to 1 second; from 1 second to 2 seconds; and from 2 seconds to 3 seconds.
      iii. What do you notice about the change in the distance of the car from the stop sign for successive *equal* changes is the number of seconds since the car started to move away from the stop sign?
   b. Is your graph concave up, concave down, or some combination of both on the interval $0 < t < 4$? Justify your answer from above.

16. For each of the following graphs, determine the intervals on which the graph is:
    i.   concave up
    ii. concave down

a.

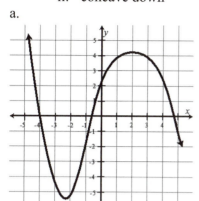

b.

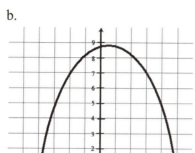

c.

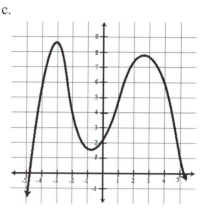

For Problems 17-18: Fire pits are frequently used to cook meat. A fire is built to heat the rocks in the bottom of the pit. Once the pit reaches the desired temperature the meat is then put in the pit and covered.

17. The table below represents the air temperature (in degrees Fahrenheit) of a pit in terms of the number of hours that have elapsed since the meat was added to the pit.

Number of hours since meat was added to the pit	Air temperature (in degrees Fahrenheit) of the pit
0	575
2	400
4	320
6	275

a. Using the table of values above:
    i.   Sketch a graph of the air temperature (in degrees Fahrenheit) of the pit in terms of the number of hours since the meat was added to the pit. (Be sure to label your axes).
    ii. Represent the changes in the air temperature of the pit from 0 hours to 2 hours; from 2 hours to 4 hours; and from 4 hours to 6 hours.
    iii. What do you notice about the change in the air temperature of the pit for successive equal change is the number of hours since the meat was added to the pit?
b. Let t represent the number of hours since the meat was added to the pit. Determine the average rate of change of the air temperature of the pit on the following time intervals.
    i. $0 \le t \le 2$
    ii. $2 \le t \le 4$
    iii. $4 \le t \le 6$
c. Describe what these average rates of change tell you about how the air temperature of the pit is changing over the time interval from $t = 0$ to $t = 6$ hours.
d. Is your graph concave up, concave down, or some combination of both on the interval $0 < t < 6$? Justify your answer from above.

18. The table below represents the internal temperature (in degrees Fahrenheit) of the meat in terms of the number of hours that have elapsed since the meat was added to the pit.

Number of hours since meat was added to the pit	Internal temperature (in degrees Fahrenheit) of the meat
0	45
1	80
3	135
10	200

   a. Using the table of values above:
      i. Sketch a graph of the internal temperature (in degrees Fahrenheit) of meat in the pit in terms of the number of hours since the meat was added to the pit. (Be sure to label your axes).
      iii. Represent the changes in the internal temperature of the meat from 0 hours to 1 hours; from 1 hours to 3 hours; and from 3 hours to 10 hours.
      iii. Using your graph from part (ii), what do you notice about the changes in the internal temperature of the meat for successive change is the number of hours since the meat was added to the pit? Do these changes tell you anything about how the internal temperature of the meat is changing over the time interval from $t = 0$ to $t = 10$ hours? If so, explain what the changes convey about the quantities in the situation. If not, explain why not.
   b. Let t represent the number of hours since the meat was added to the pit. Determine the average rate of change of the internal temperature of the meat over the following time intervals.
      i. $0 \le t \le 1$
      ii. $1 \le t \le 3$
      iii. $3 \le t \le 10$
   c. Describe what these average rates of change tell you about how the internal temperature of the meat is changing over the time interval from $t = 0$ to $t = 10$ hours.
   d. Is your graph concave up, concave down, or some combination of both on the interval $0 < t < 10$? Justify your answer from above.

19. For each of the following graphs, determine whether the rate of change of the distance from the bench with respect to the number of seconds elapsed is increasing, decreasing, or constant for successive intervals of the domain. Be sure to justify your answer.

   a.
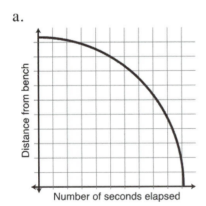

   b.
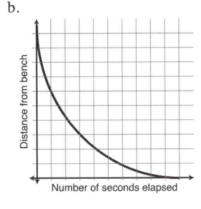

   c.
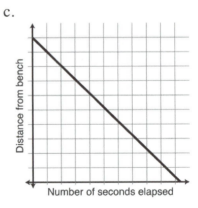

20. The graph below represents Sally's distance from her house in terms of the number of minutes t since Sally started walking.

a. Interpret the meaning of the point (20, 1950).
b. Represent the changes in Sally's distance from her house (in feet) from 0 minutes to 10 minutes; from 10 minutes to 20 minutes; and from 20 minutes to 30 minutes.
c. What do you notice about the change in Sally's distance from her house for successive equal change is the number of minutes since Sally started walking?
d. Describe what these successive changes in Sally's distance from her house for equal change in the number of minutes since Sally started walking tell you about how Sally's distance from her house is changing over the time interval from $t = 0$ to $t = 30$ minutes.
e. Describe how Sally's speed (in yards per minute) is changing as the number minutes since Sally started walking increases from 0 to 30 minutes.

21. The table of values below represents the height of the water in a bottle in terms of the volume of water in the bottle.

Volume of water in the bottle (cups)	Height of the water in the bottle (inches)
0	0
1	1.1
2	2.6
3	5.0
4	8.6
5	13.7
6	20.5
7	29.2
8	40.0

a. What is the average rate of change of the height of the water with respect to the volume of water as the volume of water increases from 2 to 3 cups? From 3 to 4 cups?
b. As the volume of water increases by equal increments, how does the average rate of change of the height of the water with respect to the volume change? *(continues...)*

c. For arbitrarily small amounts of water added to the bottle how will the average rate of change of the height of the water with respect to the volume of the water change?

d. Make a careful sketch of the bottle that could produce the height-volume relationship conveyed in the table above.

22. Watch the video titled "Co-Varying Quantities and Changing Rate of Change of Non-Linear Polynomial Functions" and answer the following questions. (The video can be found in the online textbook, Module 5 p. 11)

a. Provide an example of a context in which the values of the output quantity decrease and the average rates of change of the output quantity with respect to the input quantity are increasing, as the value of the input quantity increases.

b. Provide an example of a context in which the values of the output quantity decrease and the average rates of change of the output quantity with respect to the input quantity are decreasing, as the value of the input quantity increases.

b. For the values of r, a, n, and c given in the video is $r < n$ or is $n < r$? Explain your reasoning.

c. Is -4 greater than or less than -30? Explain your reasoning.

III. TRANSFORMATIONS OF POLYNOMIAL FUNCTIONS (TEXT: S9)

23. A coffee shop finds that its weekly profit P (measured in dollars) as a function of the price x (measured in dollars) the coffee shop charges per cup is given by the function

$$P = f(x) = -2000x^2 + 8000x - 2000$$

a. Describe the meaning of $g(x) = f(x - 1.5)$ and $h(x) = f(x) - 1.5$. Do either g or h have the same maximum profit as f? Explain.

b. Explain what $f(x+35)$ and $f(x) + 35$ represent in this context.

24. The graph of a polynomial function f is given below.

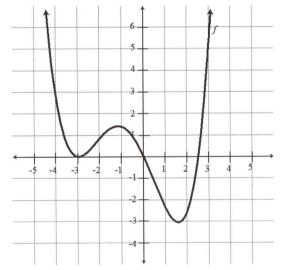

Sketch a graph of the following functions. Compare the behavior of each function below to the function f given above.

a. $g(x) = -f(x)$

b. $h(x) = f(-x)$

c. $k(x) = f(x - 2)$

d. $p(x) = f(x) - 2$

25. The graphs of polynomial functions *f* and *g* are given below.

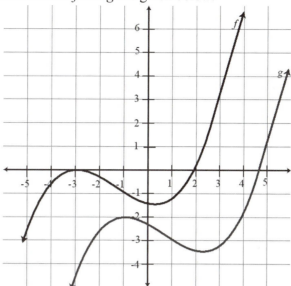

a. How do the output values of the function f compare with those of g?
b. Express the function *f* in terms of *g*.
c. Express *g* in terms of function *f*.

26. The graphs of polynomial functions *f* and *g* are given below.

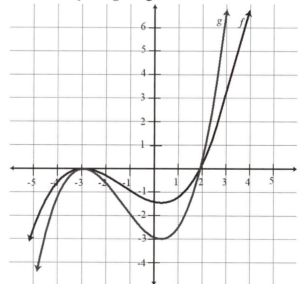

a. How do the output values of the function *f* compare with those of g?
b. How do the average rates of change of functions *f* and *g* compare on successive 1 unit intervals of the domain? Explain your reasoning?
c. Express the function *f* in terms of *g*.
d. Express *g* in terms of function *f*.

27. Each summer Primo Pizza and Pizza Supreme compete to see who has the larger summer profit. Let f be the function that determines the profit of Primo Pizza in terms of the number of days since June 1. Let g be the function that determines the profit of Pizza Supreme in terms of the number of days since June 1.
    a. Suppose Primo Pizza's profit each day is two times as large as the profit of Pizza Supreme.
        i. Express the function f in terms of the function g.
        ii. Express the function g in terms of the function f.
    b. Suppose Primo Pizza's profit each day is two hundred dollars more than the profit of Pizza Supreme.
        i. Express the function f in terms of the function g.
        ii. Express the function g in terms of the function f.
    c. Suppose Primo Pizza's profit on a given day is always the same as the profit of Pizza Supreme's profit two days later.
        i. Express the function f in terms of the function g.
        ii. Express the function g in terms of the function f.

28. The function g is defined by $g(x) = 2f(x+3) - 4$. Explain how the behavior of function g compares to the behavior of function f.

29. Factor the following expressions as much as possible.
    a. $-8x^2 - 24x$
    b. $x^2 + 6x - 16$
    c. $2x^2 + 12x + 10$
    d. $2x^2 - 5x - 12$
    e. $3x^2 - 27$
    f. $6x^2 - x - 2$

IV. QUADRATIC FUNCTIONS (TEXT: S9)

30. Consider the functions $h(x) = x^2$ and $f(x) = 2^x$.
    a. How does the growth of the quadratic function h compare to the growth of the exponential function f?
    b. Does f or h grow faster in the long run? That is, for large values of x, which of $f(x)$ or $h(x)$ will be larger?
    c. For what value(s) of x does $f(x) = g(x)$?

31. Determine the roots of the following quadratic functions.
    a. $f(x) = (3x - 2)(x + 4)$
    b. $g(x) = x^2 - 9$
    c. $h(x) = 2x^2 + 5x + 3$
    d. $k(x) = -3x^2 + 14x - 16$

32. Determine the vertex of the following quadratic functions.
    a. $g(x) = (3x - 7)(2x + 6)$
    b. $j(x) = x^2 - 16$
    c. $p(x) = 4x^2 + 8x + 3$

33. Determine the roots and axis of symmetry for the following quadratic functions.
    a. $f(x) = x^2 - 6x + 9$
    b. $g(x) = (x+3)(-2x-5)$
    c. $h(x) = -x^2 + 7x - 6$

34. Let $f(x) = x^2$, $g(x) = (x-4)^2$, $h(x) = (x-4)^2 + 7$, and $p(x) = 3(x-4)^2 + 7$.
    a. Determine the roots of the functions defined above.
    b. Determine the axis of symmetry for each of the functions defined above.
    a. Sketch a graph of each of the functions defined above.
    c. Compare the graphs (including roots, vertex, and output values) of f, h, g, and p.

35. Given a quadratic function g defined by $g(x) = ax^2 + bx + c$, and the solutions $x = 2 \pm 3.4$ to the quadratic equation $ax^2 + bx + c = 0$, determine if possible:
    a. the roots of g.
    b. the vertex of g.

36. Given a quadratic function g defined by $g(x) = ax^2 + bx + c$, and the solutions $x = -4.7 \pm 5$ to the quadratic equation $ax^2 + bx + c = 0$, determine if possible:
    a. the roots of g.
    b. the vertex of g.

37. The function f represents the height of a ball h above the ground (measured in feet) in terms of the amount of time t (measured in seconds) since the ball was thrown upward from a bridge that is some distance above the ground. $h = f(t) = -16t^2 + 48t + 120$
    a. Approximately how high is the bridge above the ground? Justify your answer.
    b. When does the ball hit the ground? Justify your answer.
    c. Construct a rough sketch of the graph of the function f that relates the height of the ball above the ground and the amount of time since the ball was thrown from the bridge.
    d. After how many seconds does the ball reach its maximum height above the ground? What is the maximum height above the ground reached by the ball? On the graph you constructed in part (c) illustrate the point that represents the ball's maximum height above the ground.
    e. On what intervals of time is the velocity of the ball increasing? Decreasing? Constant? Explain how you determined each of your responses by describing how the time since the ball was thrown and distance of the ball above the ground change together.
    f. How can your graph be used to estimate the instantaneous velocity of the ball at specific values of time since the ball was thrown? What is the velocity of the ball when it reaches its maximum height above the ground?

38. Suppose a rock is thrown upward from a bridge into a river below. The height of the rock above the surface of the water h (measured in feet) can be modeled by the function $h = f(t) = -16t^2 + 23t + 37$ where t is the number of seconds since the rock was thrown.
    a. How high is the bridge?
    b. Draw a sketch of the graph of f.
    c. Evaluate $f(0.5)$ and explain its meaning in the context of the problem.
    d. Determine the number of seconds after the rock was thrown when the rock hit the water.
    e. Determine the number of seconds after the rock was thrown when the rock reaches its maximum height above the water.

39. A rock is thrown upward from a bridge that is 20 feet above a road. The rock reaches its maximum height above the road 0.91 seconds after it is thrown and contacts the road 2.35 seconds after it was thrown. Use your knowledge of the symmetry of a parabola and the given information to develop a quadratic function that relates the time since the rock was thrown to the height of the rock above the bridge.

40. A penny is thrown from the top of a 30.48-meter building and hits the ground 3.45 seconds after it was thrown. The penny reached its maximum height above the ground 0.823 seconds after it was thrown.
    a. Define a quadratic function h that expresses the height of the penny above the ground (measured in meters) as a function of the elapsed time (measured in seconds) since the penny was thrown. (Hint: Determine the zeros of the quadratic function, and then use the height of the building to find the value of the leading coefficient.)
    b. What is the maximum height of the penny above the ground?

41. A coffee shop finds that its weekly profit P (measured in dollars) is determined by the price x (in dollars) that the coffee shop charges for a cup of coffee. The profit P can be determined by the function $P = f(t) = -4000x^2 + 12000x - 4000$
    a. Determine the maximum weekly profit and the price for a cup of coffee that produces the maximum profit.
    b. Describe the meaning of $g(x) = f(x - 2)$ and $h(x) = f(x) - 2$. Do either g or h have the same maximum profit as f? Explain. What is the price per cup of coffee that produces the maximum profit of the function f.
    c. Explain what $t(x) = f(x+60)$ and $p(x) = f(x) + 60$ represent in this context.

42. Determine if each of the following are true or false. Provide a brief explanation to justify each of your answers.
    a. A parabola has either zero or two x-intercepts.
    b. Given the general form of the parabola, $f(x) = ax^2 + bx + c$, the vertical intercept of the graph of f is the point $(0, c)$.
    c. The vertex of a parabola is the point where the graph of a quadratic function changes from increasing to decreasing, or decreasing to increasing.
    d. The zeros of a quadratic function are the horizontal intercept(s) of the graph.
    e. The domain of a quadratic function is all real numbers.
    f. The range of a quadratic function is all real numbers.
    g. If a quadratic function f is concave down on the interval $0 < x < 4$ the values of f will decrease on this interval.

43. Find the formula for a parabola (a quadratic function) that has horizontal intercepts (roots) at $x = -6$ and $x = 2$ and passes through the point $(0,5)$. What is the vertex of this parabola?

44. Find the formula for a parabola (a quadratic function) that has horizontal intercepts (roots) at $x = 1$ and $x = 3$ and passes through the point $(0, -10)$. What is the vertex of this parabola?

45. Use the quadratic formula to determine the roots of the following functions.
    a. $f(x) = x^2 + 5x$
    b. $g(x) = x^2 + 4x - 3$
    c. $k(x) = 10x^2 - 8x - 21$
    d. $m(x) = -3x^2 - 2 + 12x$
    e. $n(x) = 7 - 10x + x^2$

46. Use the quadratic formula to solve the following equations for x.
    a. $-7x^2 + 13x = -6$
    b. $3x^2 - 4x - 4 = 12$
    c. $-6x^2 + 22 = 35x$
    d. $2x^2 + 5x = 12x - 2$

47. Given the quadratic functions in factored form, write them in equivalent standard form. Finally construct a graph of each function and label its roots and maximum or minimum value.
    a. $h(x) = (2x)(3x - 4)$
    b. $m(x) = (x - 5)(2x + 3)$
    c. $f(x) = -(3x + 3)(2x - 4)$

48. Given the quadratic functions in standard form, write them in equivalent factored form. Finally construct a graph of each function and label its roots and maximum or minimum value.
    a. $p(x) = 2x^2 - 5x - 3$
    b. $h(x) = 3x^2 + 11x - 4$
    c. $k(x) = -x^2 + 9x - 14$

49. Determine the vertex of each quadratic function by completing the square. (Note that the product of completing the square only changes the form of a quadratic function. The values represented by the function are unchanged.)
    a. $f(x) = x^2 - 4x + 1$
    b. $h(x) = x^2 + 8x - 7$
    c. $k(x) = 2x^2 + 8x - 3$
    d. $p(x) = -2x^2 + 6x - 1$
    e. $s(x) = x^2 + 6x + 8$
    f. $m(x) = -x^2 + 14x + 3$
    g. $p(x) = 3x^2 + 2x + 1$

50. Given the quadratic functions in factored form, write them in equivalent standard form and then complete the square to find the function's maximum or minimum value. Finally construct a graph of each function and label its roots and maximum or minimum value.
    a. $f(x) = (2x+1)(x-5)$
    b. $h(x) = (x-7)(-3x+4)$

51. Why are second order polynomials called quadratics?

52. Watch the video titled "Rock Throw Revisited: Creating the Formula" and answer the following questions. (The video can be found in the online textbook, Module 5 p. 49)
    a. Explain how to determine the value of the root of a quadratic function given the other root and the vertex.
    b. Which of the following represents the roots of function that models the rock throwing problem discussed in the video?
        i. 0.91 ± 1.44
        ii. 2.35 ± 2
        iii. 2.35 ± 0.53
    c. Why is factored form of a function useful?
    d. Suppose the zeros of a quadratic function are $x = 2$ and $x = -0.35$ and the graph of the function passes through the point $(0, 2)$. Determine the formula for this quadratic function.

X. ROOTS AND END BEHAVIOR OF POLYNOMIAL FUNCTIONS (TEXT: S1, 6, 7, 8)

53. What is the general form of a polynomial function? Identify the constant term and leading coefficient.

54. Which of the following are polynomial functions. Justify your answer.
    a. $f(x) = 2^x$
    b. $g(x) = 5$
    c. $h(x) = \frac{5}{x} - 3x^2$
    d. $p(x) = 5x^4 + 3x^2 - 122$
    e. $r(x) = 5^{-2} - 3x$

55. Determine the roots of the following polynomial functions.
    a. $g(x) = 3x(2x-4)(x+2)^2$
    b. $h(x) = x^2(x-4)^2(x^3-8)$
    c. $h(x) = x^3 + 6x^2 + 3x$
    d. $k(x) = 2x^2 - 5x - 3$

56. Determine the roots of the following polynomial functions.
    a. $f(x) = 4x^2 + 8x + 2$
    b. $g(x) = x^3 - x^2$
    c. $s(x) = -(2x+4)(3x-1)^2(x-2)$
    d. $n(x) = 2x(x-7)(3x+4)$

57. Given the function $f(x) = (x-3)(x+1)(x-2)$
    a. Determine the roots of f.
    b. As x increases without bound, describe the behavior of $f(x)$.
    c. Explain why you can determine the behavior of f as x increases without bound by considering the sign of the output values of f to the right of its largest root.
    d. As x decreases without bound, describe the behavior of $f(x)$.
    e. Explain why you can determine the behavior of f as x increases without bound by considering the sign of the output values of f to the left of its smallest root.
    f. What is the behavior of f on the intervals $-1 < x < 2$ and $2 < x < 3$?
    g. Use your responses in part (a) – (f) to sketch a graph of the function f.

58. For the following polynomial functions determine:
    i. the interval(s) on which the output of the function is positive. Show your work.
    ii. the interval(s) on which the output of the function is negative. Show your work.
    a. $f(x) = x(3x+6)(x-1)$
    b. $g(x) = 2x^3 - 4x^2 + 4x$
    c. $h(x) = 3x(2x-5)(x+1)$

59. Define three polynomial functions that have roots at $x = 2$, $x = 4$, and $x = -3$.

60. Define three polynomial functions that have roots at $x = 1$, $x = -1$, and $x = -5$.

61. Find the formula for a polynomial function that has horizontal intercepts (roots) at $x = 2$, $x = 5$, and $x = -4$ and passes through the point $(3, 6)$.

62. For each of the polynomial functions given, identify the leading term of the function and describe the end behavior of the function based on your analysis of the leading term.
    a. $f(x) = 12x^3 - 2x^5 + 3x - 2$
    b. $g(x) = 3x(x-2)(x+4)(-2x+3)^2$
    c. $h(x) = 3x^3 - 4x + 8x^{10} - 3$
    d. $m(x) = -2(4+x)(2x-7)$

63. For each of the polynomial functions given, identify the leading term of the function and describe the end behavior of the function based on your analysis of the leading term.
    a. $f(x) = 3x(x-7)(x+2)(x-4)$
    b. $h(x) = 2x^2(-x+4)(x-7)^2$
    c. $p(x) = 2x(3x-7)(4x+1)$
    d. $s(x) = 3x^2 + 4x - 7x^8 + 6x - 2$

64. For each of the polynomial functions given, identify the leading term of the function and describe the end behavior of the function based on your analysis of the leading term.
    a. $h(x) = 2x(x+5)(x-1)^2$
    b. $k(x) = -(x+3)(x-4)^2$
    c. $g(x) = x^6 - 7x^5 + 13x^4 + 7x^3 - 34x^2 + 4x + 24$
    d. $f(x) = x^3 - 4x^2 + x + 6$

65. Given the function $f(x) = x^2(3x-4)(2x+5)^3$

    a. Determine the roots of f
    b. Describe whether the graph of f will cross through (change signs from positive to negative) or only bounce off (touch the axes and change directions) the axes at each of the roots.
    c. Evaluate $f(0)$.
    d. Examine the leading term of f to determine the end-behavior of the function.
    e. Determine on what interval(s) of the domain f is increasing.
    f. Determine on what interval(s) of the domain f is decreasing.
    g. Use your responses in part (a) – (f) to sketch a graph of f.

66. Given the function $g(x) = x^3(2x-4)^2(3x-2)(-x+1)^2$

    a. Determine the roots of g.
    b. Describe whether the graph of g will cross through (change signs from positive to negative) or only bounce off (touch the axes and change directions) the axes at each of the roots.
    c. Examine the leading term of g to determine the end-behavior of g.
    d. Determine on what interval(s) of the domain g is increasing.
    e. Determine on what interval(s) of the domain g is decreasing.
    f. Use your responses in part (a) – (f) to sketch a graph of g.

67. Given the function $m(x) = -(x-1)^2(x+2)(x-4)$

    a. Determine the roots of m.
    b. Describe whether the graph of m will cross through (change signs from positive to negative) or only bounce off (touch the axes and change directions) the axes at each of the roots.
    c. Examine the leading term of g to determine the end-behavior of m.
    d. Determine on what interval(s) of the domain m is increasing.
    e. Determine on what interval(s) of the domain m is decreasing.
    f. Use your responses in part (a) – (f) to sketch a graph of m.

For Problems 68-70 use the graph of the polynomial function f given below.

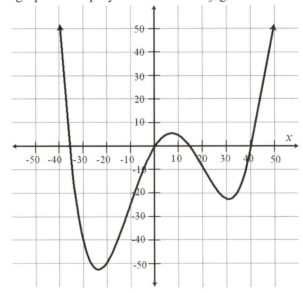

68. a. What is the domain of f?   b. What is the range of f?
    c. Evaluate $f(0)$.   d. What are the roots of f?

69. a.   On what interval(s) is f concave down?
     b.   On what interval(s) is f concave up?
     c.   On what interval(s) is f increasing?
     d.   On what interval(s) is f decreasing?

70. a.   Does the function have odd or even degree? Explain your reasoning.
     b.   Describe the behavior of f as x increases without bound.
     c.   Describe the behavior of f as x decreases without bound.
     d.   Are there any factors with an even power?

For Problems 71-73 use the graph of the polynomial function f given below.

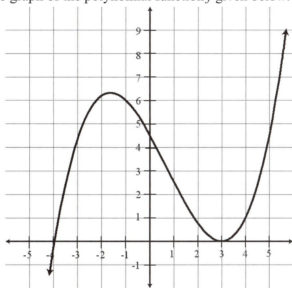

71. a.   What is the domain of f?
     b.   What is the range of f?
     c.   Evaluate $f(0)$.
     d.   What are the roots of f?

72. a.   On what interval(s) is f concave down?
     b.   On what interval(s) is f concave up?
     c.   On what interval(s) is f increasing?
     d.   On what interval(s) is f decreasing?

73. a.   Does the function have odd or even degree? Explain.
     b.   Describe the behavior of f as x increases without bound.
     c.   Describe the behavior of f as x decreases without bound.
     d.   Are there any factors with an even power?

74. Evaluate each of the following
     a. $f(x) = (8.5 - 2x)(11 - 2x)x$ when $x = 3.4$
     b. $g(x) = 87.4 - 53x$ when $x = 1.2$
     c. $h(x) = 3x^3 + 2x - 7$ when $x = -3.2$

75. Determine if each of the following statements are true or false. Provide a brief explanation to justify each of your answers.
    a. The roots of a polynomial function are rational numbers.
    b. The process of finding zeros of a polynomial function is also referenced as finding the roots of the polynomial function.
    c. We can apply the zero-product property to determine the zeros of a polynomial function in factored form.
    d. The function used to model the box problem in Module 3 has two roots, at $x = 0$ and $x = 4.25$.
    e. The zeros of any function f represent the value(s) of x that when input into the function f return a value of 1 for the output variable $f(x)$.
    f. The graph of a polynomial function is always a smooth curve.
    g. Polynomial functions are not always continuous.

76. Watch the video titled "End Behavior of Polynomial Functions" and answer the following questions. (The video can be found in the online textbook, Module 5 p. 31)
    a. How do we represent the phrase "x increases without bound"?
    b. Why is it worthwhile to consider the relative magnitude of each term in a polynomial function?
    c. True or False: When we consider the end behavior of a function we are looking at how the output values change for input values close to zero. If the statement is false, rewrite the statement so it is true.
    d. Suppose $f(x) = x^3 - 2x^2 + 5x - 100$. As x increases without bound the value of $f(x)$ is dominated by what?

1. Determine whether the following sequences of numbers are increasing or decreasing (Note: A sequence of all negative numbers is increasing if successive numbers are getting closer to 0.)

   a. $\frac{1}{100}, \frac{1}{200}, \frac{1}{320}, \frac{1}{450}, \frac{1}{1089}$

   b. $\frac{2}{100}, \frac{5}{40}, \frac{10}{35}, \frac{15}{9}, \frac{50}{6}$

   c. $\frac{100}{2}, \frac{40}{5}, \frac{35}{10}, \frac{9}{15}, \frac{6}{50}$

   d. $\frac{-1}{100}, \frac{-1}{200}, \frac{-1}{320}, \frac{-1}{450}, \frac{-1}{1089}$

A rational function is a function whose output is the relative size of one polynomial's output with respect to another polynomial's output. For example, if p and r are polynomial functions, then the function h where $h(x) = \dfrac{p(x)}{r(x)}$ is said to be a **rational function**.

---

2. The graphs of $y = p(x)$ and $y = r(x)$ are given below. Suppose the rational function h is defined by $h(x) = \dfrac{p(x)}{r(x)}$. In order to consider the behavior of the function h, let's consider the following.

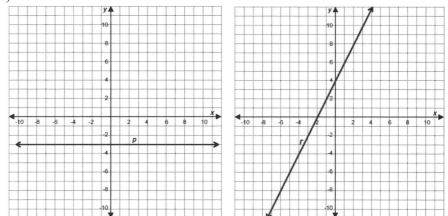

   a. As x increases to -2, describe whether the value of each function is (positive or negative) and (constant, increasing or decreasing).
      i.   The value of $p(x)$ is…
      ii. The value of $r(x)$ is…
      iii. The value of $h(x)$ is…

   b. Evaluate $h(-2)$.

   c. As x increases from -2 to 0, describe whether the value of each function is (positive or negative) and (constant, increasing or decreasing).
      i.   The value of $p(x)$ is…
      ii. The value of $r(x)$ is…
      iii. The value of $h(x)$ is…

   d. Evaluate $h(0)$.

   e. As x increases without bound from 0
      i.   The value of $p(x)$ is…
      ii. The value of $r(x)$ is…
      iii. The value of $h(x)$ is…

---

f. Sketch a graph of the function *h*.

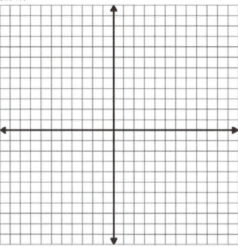

3. Suppose $h(x) = \dfrac{p(x)}{r(x)}$ where *p* and *r* are polynomial functions.

    a. For what value(s) of *x* is *h*(*x*) undefined?   b. For what value(s) of *x* does *h*(*x*) = 0?

4. A small airport is considering selling jet fuel at its airport. There is an initial investment of $3,200,000 to install tanks for the fuel. The price the airport pays for fuel is $3.90 per gallon. The airport owner has asked us to develop a mathematical model to explore how the average cost per gallon of his investment (including the initial investment) varies with the number of gallons of fuel purchased *x*.

    a. As a first step, define a function *f* that describes how the cost paid by the airport varies with the number of gallons of fuel, *x*, purchased.

    b. The average cost per gallon of the investment involves determining the relative size the cost *f*(*x*) and the number of gallons purchased *x*.

    Use the following table to determine the average cost per gallon to the airport for supplying the corresponding number of gallons of fuel. (Note: we are going to focus our attention on what happens for relatively small and relatively large changes in input throughout this investigation.)

x	Cost *f*(*x*)	Average Cost *g*(*x*)
0.01		
0.1		
0.5		
1		
10		
100		
1,000		

    c.   Define a function g that relates the airport's average cost per gallon (measured in dollars) as a function of the number of gallons of fuel supplied.

    d.   When x is positive and getting very close to 0 (which we write as $x \to 0^+$), the average cost per gallon for fuel gets very, very large. Why does this happen?

    e.   Does the average cost per gallon have a maximum value? Explain.

5. Five gallons of liquid flavoring are poured into a large vat. Water will be added to the vat and mixed with the flavoring to produce a drink that will be bottled and sold.
    a.   Suppose water is added until the total mixture is 7 gallons. What is the ratio of flavoring to water?

    b.   Suppose water is added until the total mixture is 18 gallons. What is the ratio of flavoring to water?

    c.   Define a function f with the total volume of the mixture (in gallons), x, as the independent quantity, that determines the ratio of flavoring to water R in the mixture. What is the practical domain for this function?

    d.   What does it mean to say "as $x \to 5^+$"?

    e.   What happens to the value of f as $x \to 5^+$? Why does this make sense?

6. Given that $f(x) = \dfrac{1}{x-8}$, answer the questions below.
    a.   Complete the following table of values. Show the calculations that provided your answers.

x	$f(x)$
7	
7.5	
7.9	
7.99	
7.999	

    b.   What do you notice as x approaches 8 from the left (from values less than 8, written $x \to 8^-$)? Why does this happen?

c.   Complete the following table of values. Show the calculations that provided your answers.

x	$f(x)$
9	
8.5	
8.1	
8.01	
8.001	

d.   What do you notice as x approaches 8 from the right (from values greater than 8, written $x \to 8^+$)? Why does this happen?

e.   Using a calculator, graph f. Explain how the graph supports your conclusions above.

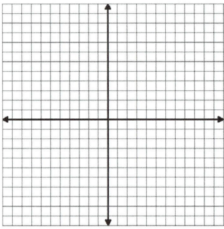

7.   Given that $f(x) = \dfrac{-3}{x+4}$, answer the questions below.

a.   Complete the following table of values.

x	$f(x)$
−4.1	
−4.01	
−4.001	
−3.999	
−3.99	
−3.9	

b.   What do you notice as x approaches −4 from the left ($x \to -4^-$)? Why does this happen?

c. What do you notice as x approaches 8 from the right ($x \rightarrow -4^+$)? Why does this happen?

d. Using a calculator, graph f. Explain how the graph supports your conclusions in parts (b) and (d).

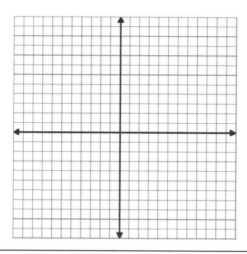

A vertical asymptote occurs at $x = a$ (a is a real number) if
- as $x \rightarrow a^-$ (read, as x approaches a from the negative) $f(x)$ increases or decreases without bound

and
- as $x \rightarrow a^+$ (read, as x approaches a from the positive) $f(x)$ increases or decreases without bound

8. For each of the following functions, predict where a vertical asymptote will exist, then test your prediction using a graphing calculator.

a. $f(x) = \dfrac{8}{x+1}$

b. $g(x) = \dfrac{x}{2x-8}$

c. $h(x) = \dfrac{6x^2 - 24}{2(x-3)(x+1)}$

d. $m(x) = \dfrac{x+3}{x^2 + 4x - 5}$

e. $n(x) = \dfrac{x^5 + 1}{x^2 + 1}$

f. $p(x) = \dfrac{2(x+6)}{x+6}$

9. Write an explanation describing how to determine the vertical asymptotes of a rational function and *why* this approach works.

4. Given that $f(x) = \dfrac{1}{x-8}$, answer the questions below.

   a. Complete the following table of values.

x	$x-8$	$f(x) = \dfrac{1}{x-8}$
100		
10,000		
1,000,000		

   b. How do the output values of f change as x increases without bound? Why does this happen?

   c. Complete the following table of values.

x	$x-8$	$f(x) = \dfrac{1}{x-8}$
−100		
−10,000		
−1,000,000		

   d. How do the output values of f change as x decreases without bound (written $x \to -\infty$)? Why does this happen?

   e. Using a calculator, graph f. Explain how the graph supports your conclusions above.

5. Given that $f(x) = \dfrac{3x}{x+4}$, answer the questions below.

   a. Complete the following table of values.

x	$3x$	$x+4$	$f(x) = \dfrac{3x}{x+4}$
$-1{,}000{,}000$			
$-10{,}000$			
-100			
100			
$10{,}000$			
$1{,}000{,}000$			

   b. What do you notice as x increases without bound? Why does this happen?

   c. What do you notice as x decreases without bound? Why does this happen?

   d. Using a calculator, graph f. Explain how the graph supports your conclusions in parts (b) and (c) above.

A *horizontal asymptote* is a horizontal line that we draw at a value of the output when the following behavior occurs: Let a be a real number. If $f(x) \to a$ as x increases without bound and/or as x decreases without bound , then f has a horizontal asymptote at $y = a$. In Problem #1 above, a horizontal asymptote occurred at $C = 3.90$. In Problem #2, a horizontal asymptote occurred at $R = 0$.

6. Where did the horizontal asymptotes occur in Problems 3, 4, and 5?

7. There are two ways to determine horizontal asymptotes. The first method is to use a table or graph to examine what happens to the output values of a function as x increases and decreases without bound. A second method uses the formula of the function and compares the long-run behavior of the polynomial functions that make up the numerator and denominator of a rational function. The basic reasoning goes like this. Consider the polynomial function $p(x) = x + 5$. When the magnitude of x gets very large, the "+ 5" portion of the function rule is a proportionally smaller and smaller part of determining the final value of the expression $x + 5$. For example, when $x = 20$, $p(20) = 25$. When $x = 1,000$, $p(1,000) = 1,005$. When $x = 100,000,000$, $p(100,000,000) = 100,000,005$. We can say that as x increases without bound, $p(x) \to x$ (and likewise as x decreases without bound, $p(x) \to x$) because the relative size of $x + 5$ and x are nearly identical.

   a. As x increases or decreases without bound, what happens to the value of the expression $x^2 - 10$? Why does this happen?

   b. As x increases or decreases without bound, what happens to the value of the expression $x^3 - 4x^2$? Why does this happen?

   c. As x increases or decreases without bound, what happens to the value of the expression $5x^2 + 12x$? Why does this happen?

   d. As x increases or decreases without bound, what happens to the value of the expression $-7x^4 + 4x^3 - x^2 + 3x + 1$? Why does this happen?

8. Using the reasoning from above, consider the function $f(x) = \dfrac{5x + 2}{x^2 - x}$. As x increases or decreases without bound, we have $(5x + 2) \to 5x$ and $(x^2 - x) \to x^2$. Therefore, $\left(f(x) = \dfrac{5x + 2}{x^2 - x} \right) \to \dfrac{5x}{x^2} = \dfrac{5}{x}$ as x increases or decreases without bound. If $f(x) \to \dfrac{5}{x}$ as x increases or decreases without bound, then we reason that there must be a horizontal asymptote at $y = 0$ since $g(x) = \dfrac{5}{x}$ has a horizontal asymptote at $y = 0$. Use this reasoning to find the horizontal asymptotes for the following functions.

   a. $f(x) = \dfrac{8}{x + 1}$

   b. $g(x) = \dfrac{x}{2x - 8}$

   c. $h(x) = \dfrac{6x^2 - 24}{2(x - 3)(x + 1)}$

d. $m(x) = \dfrac{x+3}{x^2+4x-5}$ e. $n(x) = \dfrac{x^5+1}{x^2+1}$ f. $p(x) = \dfrac{2(x+6)}{x+6}$

9. Generalizing the reasoning above, any polynomial expression
$p(x) = a_n x^n + a_{n-1} x^{n-1} + a_{n-2} x^{n-2} + \ldots + a_1 x + a_0$ will behave increasingly like $a_n x^n$ as x increases or decreases without bound. Therefore, a rational function
$r(x) = \dfrac{p(x)}{q(x)} = \dfrac{a_n x^n + a_{n-1} x^{n-1} + a_{n-2} x^{n-2} + \ldots + a_1 x + a_0}{b_m x^m + b_{m-1} x^{m-1} + b_{m-2} x^{m-2} + \ldots + b_1 x + b_0}$ will behave more and more like $\dfrac{a_n x^n}{b_m x^m}$ as x increases or decreases without bound.

a. What conditions are necessary for a rational function to have a horizontal asymptote at $y = 0$?

b. What conditions are necessary for a rational function not to have a horizontal asymptote?

c. What conditions are necessary for a rational function to have a horizontal asymptote at $y = a$ where $a \neq 0$? How can you determine the exact location of the horizontal asymptote?

Use these four rational functions (i-iv) to answer Questions 1-7.

i) $f(x) = \dfrac{x-3}{x+2}$ ii) $g(x) = \dfrac{3x^2}{(x-1)(x-3)}$ iii) $h(x) = \dfrac{x^2+1}{x-2}$ iv) $k(x) = \dfrac{5x}{x^2-4}$

1. a. Find the real zeros (or roots) of each function.

     i)   ii)   iii)   iv)

    b. Describe your method for determining the zeros of each function and explain why this method works.

2. a. Determine the vertical intercept of each function.

     i)   ii)   iii)   iv)

    b. Describe your method for determining the vertical intercept of each function and explain why this method works.

3. a. What is the domain of each function?

     i)   ii)   iii)   iv)

    b. What happens at the values you excluded from your domain in part (a)?

4. a. Using what you learned in Investigation 2, fill in the blanks below.

    i) As $x \to \infty, f(x) \to$ _____   ii) As $x \to \infty, g(x) \to$ _____
       As $x \to -\infty, f(x) \to$ _____   As $x \to -\infty, g(x) \to$ _____
       Horizontal Asymptote: _____   Horizontal Asymptote: _____

    iii) As $x \to \infty, h(x) \to$ _____   iv) As $x \to \infty, k(x) \to$ _____
        As $x \to -\infty, h(x) \to$ _____   As $x \to -\infty, k(x) \to$ _____
        Horizontal Asymptote: _____   Horizontal Asymptote: _____

b. Combine the information from your answers above to sketch a graph of each function. Once you have sketched your graphs, use a calculator to graph the functions and compare. If your sketch was incorrect, write a brief description of your error and why it was incorrect. *Note: Remember that if you ever get stuck regarding how a function behaves on a given interval, you can always create a table of values.*

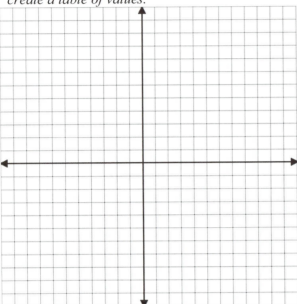

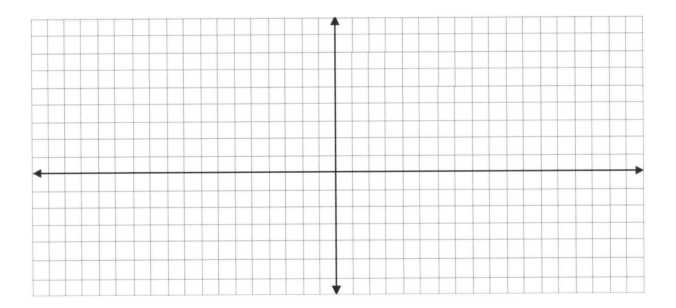

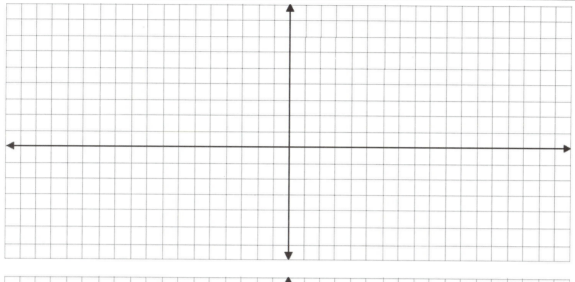

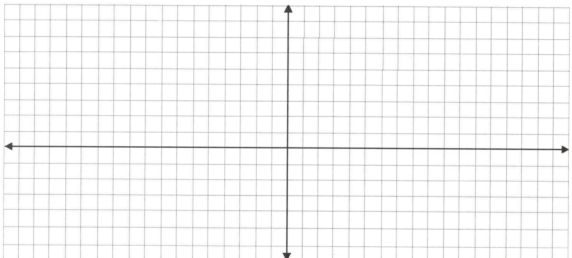

5. Sketch a graph of the following rational functions without using a graphing calculator by first finding the location of the horizontal and vertical intercepts and the horizontal and vertical asymptotes. *Note: Remember that if you need additional information about a function's behavior over some specific part of the domain you can always create a table of values to help you.*

a. $g(x) = \dfrac{x+7}{x-5}$

b. $h(x) = \dfrac{x^2-9}{x+4}$

c. $f(x) = \dfrac{2(x-3)(x-1)}{(x-4)(x+1)}$

a.

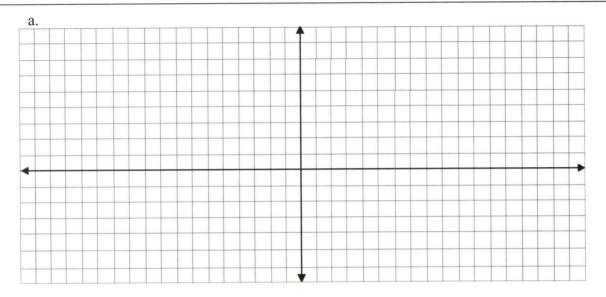

b.

c.

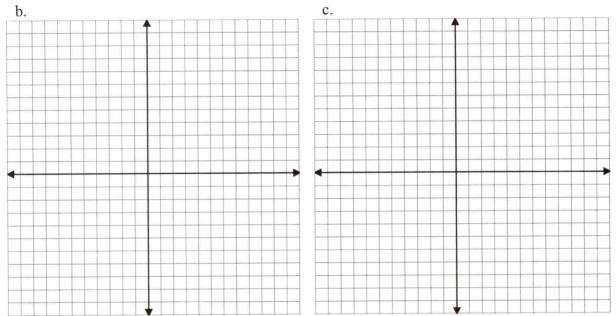

6. *Limit notation* is a concise way to communicate behavior about a function. For example, consider the function $f(x) = \dfrac{x-3}{x+2}$. We say that the *limit* of a function is the value (or $\pm\infty$) that the outputs of a function approach as the input approaches a specified value (or increases or decreases without bound). For example, we know that *f* has a horizontal asymptote at $y = 1$, so as $x \to \infty$, $f(x) \to 1$. Therefore, we say that *the limit of $f(x)$ is 1 as $x \to \infty$*, and we write it as $\lim\limits_{x \to \infty} f(x) = 1$. Similarly, since $f(x) \to 1$ as $x \to -\infty$, we say that $\lim\limits_{x \to -\infty} f(x) = 1$.

We also know there is a vertical asymptote at $x = -2$, and by plugging in a few values or looking at a graph we can determine that $f(x) \to -\infty$ as $x \to -2^-$. Therefore, we can condense $f(x) \to -\infty$ as $x \to -2^-$ using limit notation by writing $\lim_{x \to -2^-} f(x) = -\infty$. Similarly, $f(x) \to \infty$ as $x \to -2^+$ so $\lim_{x \to -2^+} f(x) = \infty$.

Use the following functions to answer the questions below. (Note: If a value does not exist, write DNE.)

$$f(x) = \frac{x+3}{x+6} \qquad g(x) = \frac{3x^2}{(x-1)(x-3)} \qquad h(x) = \frac{x^2+1}{x-2} \qquad k(x) = \frac{5x}{x^2-4}$$

a. $\lim_{x \to \infty} f(x) = \underline{\hspace{1cm}}$

$\lim_{x \to -\infty} f(x) = \underline{\hspace{1cm}}$

$\lim_{x \to -6^-} f(x) = \underline{\hspace{1cm}}$

$\lim_{x \to -6^+} f(x) = \underline{\hspace{1cm}}$

b. $\lim_{x \to \infty} g(x) = \underline{\hspace{1cm}}$

$\lim_{x \to 1^-} g(x) = \underline{\hspace{1cm}}$

$\lim_{x \to 1^+} g(x) = \underline{\hspace{1cm}}$

$\lim_{x \to 3^+} g(x) = \underline{\hspace{1cm}}$

c. $\lim_{x \to \infty} h(x) = \underline{\hspace{1cm}}$

$\lim_{x \to -\infty} h(x) = \underline{\hspace{1cm}}$

$\lim_{x \to 2^-} h(x) = \underline{\hspace{1cm}}$

$\lim_{x \to 2^+} h(x) = \underline{\hspace{1cm}}$

d. $\lim_{x \to -\infty} k(x) = \underline{\hspace{1cm}}$

$\lim_{x \to -2^-} k(x) = \underline{\hspace{1cm}}$

$\lim_{x \to 2^-} k(x) = \underline{\hspace{1cm}}$

$\lim_{x \to 2^+} k(x) = \underline{\hspace{1cm}}$

7. In each part below, you are given information about a rational function. Use the information to sketch a possible graph of the function.

a. $\lim_{x \to \infty} f(x) = -3$

$\lim_{x \to -\infty} f(x) = -3$

$\lim_{x \to 2^-} f(x) = \infty$

$\lim_{x \to 2^+} f(x) = -\infty$

b. $\lim_{x \to \infty} f(x)$ DNE

$\lim_{x \to \infty} f(x)$ DNE

$\lim_{x \to -4^-} f(x) = -\infty$

$\lim_{x \to -4^+} f(x) = \infty$

c. $\lim_{x \to \infty} f(x) = 2$

$\lim_{x \to -\infty} f(x) = 2$

$\lim_{x \to -3^-} f(x) = \infty$

$\lim_{x \to -3^+} f(x) = -\infty$

$\lim_{x \to 1^-} f(x) = -\infty$

$\lim_{x \to 1^+} f(x) = \infty$

a.

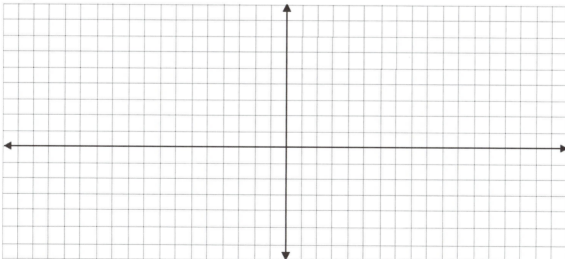

b.
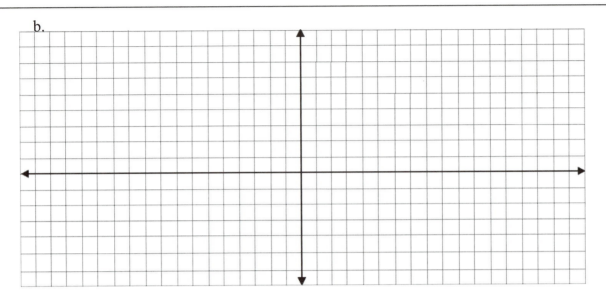

c.
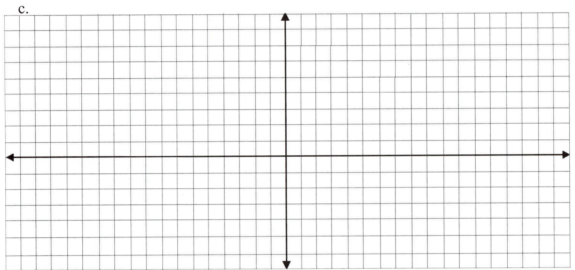

1. Use the graph of f below to answer the questions that follow.

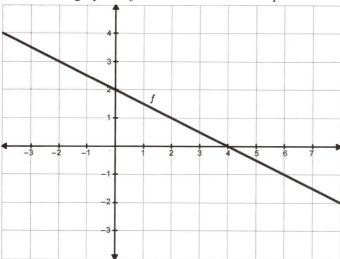

    a. Describe the important characteristics of f.

    b. Define a new function g by $g(x) = \dfrac{1}{f(x)}$.

        i. What happens to the value of $g(x)$ as $x \to 4^-$?

        ii. What happens to the value of $g(x)$ as $x \to 4^+$?

        iii. What happens to the value of $g(x)$ as x increases without bound?

        iv. What happens to the value of $g(x)$ as x decreases without bound?

        v. Sketch a graph of g.

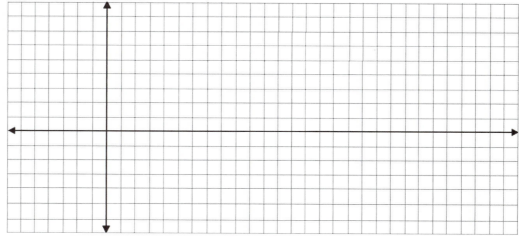

2. What, if anything, would change if the definition of g were $g(x) = \dfrac{10}{f(x)}$ instead of $g(x) = \dfrac{1}{f(x)}$?
    Explain your reasoning.

3. What, if anything, would change if the definition of g were $g(x) = \dfrac{1}{f(x)+1}$ instead of
    $g(x) = \dfrac{1}{f(x)}$? Explain your reasoning.

4. Given the graph of f below, sketch the graph of g if $g(x) = \dfrac{5}{f(x)}$.

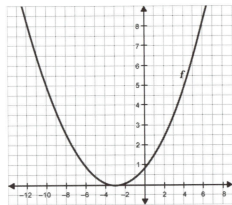

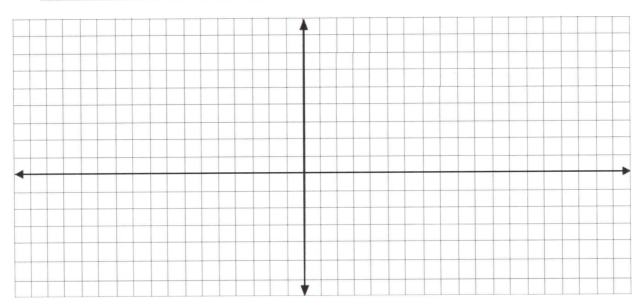

5. Given the graphs of f and g below, sketch the graph of h if $h(x) = \dfrac{f(x)}{g(x)}$.

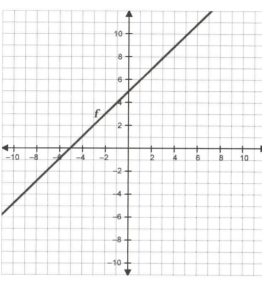
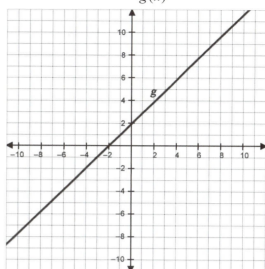

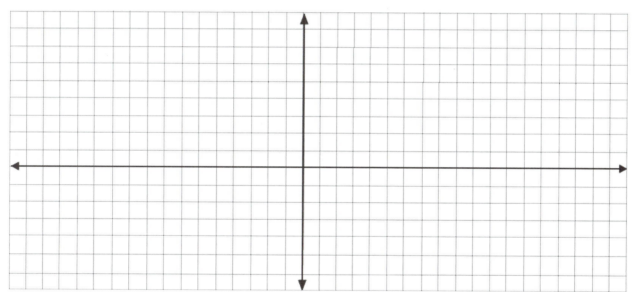

I. VERTICAL ASYMPTOTES OF RATIONAL FUNCTIONS (Text: S1, 2, 4, 6)

1. A weight loss center is so confident in their program that they created a special pricing plan. Joining the program costs a one-time fee of $49.99, and members don't have to pay anything more unless they lose weight. However, members pay the center $2.99 per pound they lose while in the program.
    a. Define a function f to express the cost C for someone joining the program and losing x pounds. What is the practical domain of this function?
    b. Suppose we are interested in determining the average cost per pound lost for someone in the program. (Note that average cost is NOT an *average rate of change*.) Describe how to determine the average cost per pound lost when x pounds are lost by a member in the program. (e.g., Describe the calculation you would perform to determine the average cost and why you would perform this calculation.)
    c. Use the following table to determine the average cost per pound lost for a member losing x pounds while in the program.

x	Cost $f(x)$	Average Cost $g(x)$
0.01		
0.1		
1		
10		
100		

    d. Define a function g that relates the member's average cost per pound lost (measured in dollars) as a function of the number of pounds lost.
    e. When x is positive and getting very close to 0 (which we write as $x \to 0^+$), what happens to the average cost per pound? Why does this happen?
    f. Does the average cost per pound have a maximum value? Explain.

2. A beverage company has just completed brewing a large batch of tea (1500 gallons). They will add sweetening to the tea, then bottle it and prepare it for distribution.
    a. Suppose 10 gallons of corn syrup is added to the mixture as a sweetener. What is the ratio of tea to corn syrup in the mixture? What if 50 gallons of corn syrup is added instead? 200 gallons?
    b. Define a function f whose input x is the amount of corn syrup added (in gallons) and whose output R is the ratio of tea to corn syrup in the mixture. What is the practical domain for this function?
    c. Complete the following table of values for this function.

x	$R = f(x)$
0.001	
0.01	
0.1	
1	

    d. What happens to R as $x \to 0^+$? Why does this make sense?

3. Orange juice concentrate is made by taking the juice from oranges and removing the water. Consumers can purchase the concentrate and add water to produce normal orange juice. While each brand will suggest the amount of water to add to the concentrate, consumers can vary this amount to produce stronger or weaker orange juice depending on personal taste. Jermaine dumps one can of frozen orange juice concentrate into a pitcher.
   a. Suppose Jermaine adds water to the concentrate until there are three cans worth of liquid. What is the ratio of concentrate to water? What if he adds water until there are four cans of liquid?
   b. Define a function f whose input x is the total volume of the mixture (in cans) and whose output R is the ratio of concentrate to water in the mixture. What is the practical domain for this function?
   c. Complete the following table of values for this function.

x	$R = f(x)$
1.01	
1.1	
1.5	
2	

   d. What happens to R as $x \to 1^+$? Why does this make sense?

4. A cylindrical container is being designed to carry liquid that must be insulated (see diagram below). The interior cylinder will hold the liquid, and the insulation must be 2 inches thick.

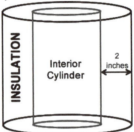

   a. If the radius of the entire container (interior cylinder and insulation) is 4.5 inches, what is the radius of the interior cylinder? What if the radius of the entire container is 5.9 inches? x inches?
   b. The formula for the volume of a cylinder is given by $V = \pi r^2 h$. Suppose the container must hold exactly 200 in³ of liquid. How tall must the container be if the radius of the entire container is 4.5 inches and the interior cylinder holds exactly 200 in³? What if the radius of the entire container is 5.9 inches?
   c. Define a function f whose input x is the radius of the entire package (in inches) and whose output h is the height of the package (in inches) necessary so that the interior container holds 200 in³ of liquid.

   d. What happens to the value of h as $x \to 2^+$? What does this mean in the context of this problem?

5. The wildlife game commission introduced a herd of 5 antelope into a 20-acre wildlife refuge. The function h defines the number of antelope, N, in the herd as a function of time t (measured in years) since the herd was introduced.
$$N = h(t) = \frac{6t + 5}{0.5t + 1}$$
   a. Create a graph of h and label the axes.

b. Does the function h have a vertical asymptote? If so, find the vertical asymptote and if possible explain its meaning in the context of the problem. If it has no meaning in the problem context explain why not.

c. Find the vertical intercept and describe what it represents in the context of the problem.

6. Given that $f(x) = \dfrac{x}{x-9}$, answer the questions below.

   a. Complete the following tables of values. Show the calculations that provided your answers.

x	$f(x)$
8	
8.9	
8.99	
8.999	

x	$f(x)$
9.001	
9.01	
9.1	
10	

   b. How do the output values of f change as $x \to 9^-$? Why does this happen?

   c. How do the output values of f change as $x \to 9^+$? Why does this happen?

   d. Using a calculator, graph f. Explain how the graph supports your conclusions above.

   e. What changes if the definition of f becomes $f(x) = \dfrac{x}{9-x}$? Why does this happen?

   f. What changes if the definition of f becomes $f(x) = \dfrac{x}{x+9}$? Why does this happen?

7. Given that $f(x) = \dfrac{2}{x+7}$, answer the questions below.

   a. How do the output values of f change as $x \to -7^-$?

   b. How do the output values of f change as $x \to -7^+$?

   c. Using a calculator, graph f. Explain how the graph supports your conclusions in parts (a) and (b) above.

   d. What changes if the definition of f becomes $f(x) = -\dfrac{2}{x+7}$? Why does this happen?

8. Given that $g(x) = \dfrac{5x}{(x+1)(x-3)}$, answer the questions below.

   a. How do the output values of g change as $x \to -1^-$? As $x \to -1^+$?

   b. How do the output values of g change as $x \to 3^-$? As $x \to 3^+$?

   c. Using a calculator, graph f. Explain how the graph supports your conclusions in parts (a) and (b) above.

9. For each of the following functions, predict where a vertical asymptote will exist, then test your prediction using a graphing calculator.

   a. $f(x) = \dfrac{x-10}{x+10}$ b. $g(x) = \dfrac{x^2}{5x-9}$ c. $h(x) = \dfrac{x}{x^2+2}$

   d. $m(x) = \dfrac{6}{x^2-1}$ e. $n(x) = \dfrac{5x+5}{x+1}$ f. $p(x) = \dfrac{6-x}{(x-4)(x-5)}$

II. END BEHAVIOR OF RATIONAL FUNCTIONS (Text: S1, 2, 3, 5)

10. A beverage company has just completed brewing a large batch of tea (1500 gallons). They will add corn syrup to the tea, then bottle it and prepare it for distribution. The function f relates the ratio R of tea to corn syrup in the beverage when x gallons of corn syrup are added and is defined by

$$R = f(x) = \frac{1500}{x}.$$

a. Complete the following table of values for this function.

x	$R = f(x)$
100	
10,000	
1,000,000	
10,000,000	

b. What happens to R as x increases without bound?

c. Is it practical in this context to allow x to increase without bound? Explain.

11. The wildlife game commission introduced a herd of 5 antelope into a 20-acre wildlife refuge. The function h defines the number of antelope, N, in the herd as a function of time t (measured in years) since the herd was introduced.

$$N = h(t) = \frac{6t + 5}{0.5t + 1}$$

a. Create a graph of h and label the axes.

b. Describe the number of antelope in the herd as the number of years gets larger and larger.

c. According to the model, does the antelope population have a maximum value? If so, what is the largest population? If not, why not?

12. A cylindrical container is being designed to carry liquid that must be insulated (see diagram below). The interior cylinder will hold the liquid, and the insulation must be 2 inches thick. Function f is

defined by $h = f(x) = \dfrac{200}{\pi(x-2)^2}$ where x is the radius of the entire package (in inches) and h is the

height of the package (in inches) necessary so that the interior container holds 200 in^3 of liquid.

a. Complete the following table of values.

x	$h = f(x)$
10	
100	
1,000	
10,000	

b.   What happens to the value of h as x increases without bound? What does this mean in the context of this problem?

13. Clark's Soda Company incurred a start-up cost of $1276 for equipment to produce a new soda flavor. The cost of producing the drink is $0.26 per can. The company generates revenue by selling the soda for $0.75 per can. (Because the demand for the new soda is high, the Clark's Soda Company is able to sell every can produced.)
    a.   Define a function f to determine the cost (including the start-up cost) measured in dollars to produce x cans of soda.
    b.   Define a function g to determine the revenue (measured in dollars) generated from selling x cans of soda.
    c.   Define a function h to determine the profit (measured in dollars) from selling x cans of soda (recall that profit = revenue – cost).
    d.   Define a function A to determine the average cost per can of producing the soda.
    e.   How does A change as the number of cans produced gets larger and larger? Will the average cost function A ever reach a minimum value? Explain.

14. A salt cell is cleaned using a mixture of water and acid. There are currently 10 liters of water in a bucket. A technician adds varying amounts of liters of acid to the 10 liters of water.
    Let x = the number of liters of acid added to the 10 liters of water
    Let A = the level of acidity of the mixture (or the percentage of the solution that is acid)
    a.   Complete the following table of the acidity of the mixture (*measured in a percentage*) as acid is added to the 10 liters of water:

liters of acid, x	0	0.5	1.0	1.5	2.0	2.5	3.0	3.5
Acidity of mixture, A								

    b.   Define a function to determine the acidity of the mixture in terms of the number of liters of acid x that has been added.
    c.   Use a calculator to graph f.
    d.   Describe how the acidity of the mixture changes as the number of liters of acid added to the solution approaches positive infinity.
    e.   Is it possible for the mixture to ever reach 100% acid? Explain.
    f.   If the water is too acidic, it can damage the salt cell. If the water is not acidic enough, the mixture will not be practical in cleaning the cell. It turns out the mixture is most efficient if the acidity is between 40% and 60%. What is the range of liters of acid that should be added to the bucket of water so that the mixture is most efficient?
    g.   Define a function that accepts the desired percentage of acid as input and determines the number of liters of acid that should be added to the water as output.

15. A large mosquito repellent manufacturer produces a repellent using DEET (the most common active ingredient in insect repellents) and a moisturizer lotion. A tank for mixing the repellent is filled with a steady flow of the lotion and DEET. Soon after the filling began, the mixture was sampled and it was determined that there were 97 ounces of lotion and 4 ounces of DEET in the tank (let the time of the first sample be $t = 0$ where time is measured in a number of seconds). Additional readings revealed that the lotion was flowing in at a constant rate of 12 ounces per second and the DEET was flowing in

at a constant rate of 6.4 ounces per second. The company needs to monitor the percentage of DEET in their lotion.

a. Define a function f to determine the number of ounces of DEET in the tank as a function of time (measured in seconds).

b. Define a function g to determine the *total* number of ounces of the mixture (both lotion and DEET) that are in the tank as a function of time (measured in seconds).

c. Define a function h to determine the percentage of DEET in the mixture as a function of time (measured in seconds).

d. Fill out the table below. Round your answers to the nearest ten-thousandths.

Number of seconds elapsed, t	Percentage of DEET in the mixture, $h(t)$
0	
50	
100	
200	
500	

e. Describe how the percentage of DEET in the solution changes as time increases. Will the percentage of DEET in the mixture ever reach a maximum? Explain.

f. Describe the end-behavior (as t increases without bound) of h.

g. Use a calculator to graph h, then explain how the graph supports your answer in part (f).

16. For each function below, state the horizontal asymptote (if one exists).

a. $f(x) = \dfrac{x}{2x-3}$

b. $g(x) = \dfrac{x^2}{x+5}$

c. $h(x) = \dfrac{4x+1}{2x-10}$

d. $m(x) = \dfrac{x^2-2}{x^2+3x+2}$

e. $n(x) = \dfrac{17x+200}{10x^3-100x^2}$

f. $p(x) = \dfrac{5x^2}{x(x-4)}$

g. $q(x) = \dfrac{2x^3+7}{5x^4-10}$

h. $r(x) = \dfrac{4x^2+x+11}{(3x+1)(2x-3)}$

i. $w(x) = \dfrac{(x+9)(x-5)}{(x+6)(x+2)(x-3)}$

III. GRAPHING RATIONAL FUNCTIONS AND UNDERSTANDING LIMITS (Text: S3, 4, 5, 6)

17. Which of the following best describes the behavior of the function f defined by $f(x) = \dfrac{1}{(x-2)^2}$?

Provide a rationale for your answer.

a. As the value of x approaches positive infinity, the value of f decreases without bound.

b. As the value of x approaches positive infinity, the value of f increases without bound.

c. As the value of x approaches positive infinity, the value of f approaches 0.

d. As the value of x approaches 2, the value of f approaches 0.

e. (a) and (c)

18. For each of the following functions, identify the x-intercepts (roots), y-intercept, horizontal asymptotes, vertical asymptotes, and the function's domain. (State DNE in cases when an intercept or asymptote does not exist.) Use this information to sketch a graph of the function.

   a. $a(x) = \dfrac{3}{x-7}$

   b. $b(x) = \dfrac{x}{2x+6}$

   c. $d(x) = \dfrac{9x}{3x+3}$

   d. $f(x) = \dfrac{-x+3}{2x-1}$

   e. $g(x) = \dfrac{9x^2-144}{x^2-1}$

   f. $h(x) = \dfrac{14x}{3x-4}$

   g. $k(x) = \dfrac{x-11}{x^2-5x+6}$

   h. $p(x) = \dfrac{x(x+2)}{4x+1}$

   i. $q(x) = \dfrac{(x^2+2x-3)}{x^2-1}$

For Problems 19 and 20, use the following graph of f.

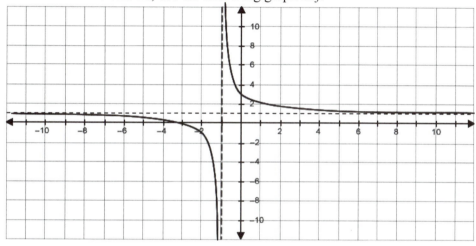

19. Use the graph of f to determine the following limits. If the limit does not exist, write DNE.

   e. $\lim\limits_{x\to\infty} f(x)$

   b. $\lim\limits_{x\to-\infty} f(x)$

   c. $\lim\limits_{x\to 2^+} f(x)$

   d. $\lim\limits_{x\to 2^-} f(x)$

   e. $\lim\limits_{x\to 2} f(x)$

   f. $\lim\limits_{x\to -1^+} f(x)$

   g. $\lim\limits_{x\to -1^-} f(x)$

   h. $\lim\limits_{x\to -1} f(x)$

20. Using the graph of f, identify the vertical and horizontal intercepts, the vertical and horizontal asymptotes, and the domain and range of the function. Then, determine a rule for the function f.

For Problems 21 and 22, use the following graph of *g*.

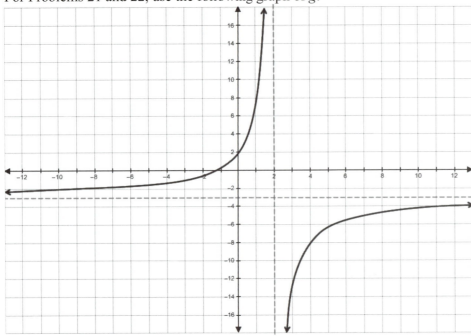

21. Use the graph of *g* to determine the following limits. If the limit does not exist, write DNE.

    f. $\displaystyle\lim_{x \to \infty} g(x)$ b. $\displaystyle\lim_{x \to -\infty} g(x)$ c. $\displaystyle\lim_{x \to 2^+} g(x)$ d. $\displaystyle\lim_{x \to 2^-} g(x)$

    e. $\displaystyle\lim_{x \to 2} g(x)$ f. $\displaystyle\lim_{x \to 6^+} g(x)$ g. $\displaystyle\lim_{x \to 6^-} g(x)$ h. $\displaystyle\lim_{x \to 6} g(x)$

22. Use the graph of *g* to identify the vertical and horizontal intercepts, the vertical and horizontal asymptotes, and the domain and range of the function. Then, determine a rule for the function *g* (i.e., represent the function algebraically).

For Problems 23 and 24, use the following graph of *h*.

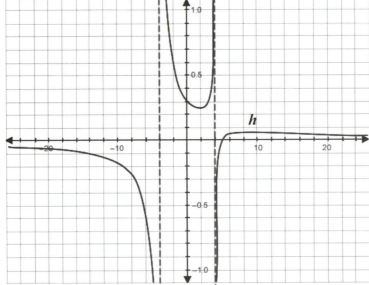

23. Use the graph of h to determine the following limits. If the limit does not exist, write DNE.

    a. $\lim\limits_{x\to\infty} h(x)$ b. $\lim\limits_{x\to-\infty} h(x)$ c. $\lim\limits_{x\to-4^+} h(x)$ d. $\lim\limits_{x\to-4^-} h(x)$

    e. $\lim\limits_{x\to-4} h(x)$ f. $\lim\limits_{x\to4^+} h(x)$ g. $\lim\limits_{x\to4^-} h(x)$ h. $\lim\limits_{x\to4} h(x)$

24. Use the graph of h to identify the vertical and horizontal intercepts, the vertical and horizontal asymptotes, and the domain and range of the function. Then, determine a rule for the function h.

For Problems 25 and 26 using the following table of values for the rational functions f, g, and h.

x	-10000	-1000	-100	100	1000	10000
$f(x)$	-10001.9997	-1001.9970	-101.9694	97.9706	997.9970	9997.9997
$g(x)$	-6.0004	-6.0040	-6.0404	-5.9604	-5.9960	-5.9996
$h(x)$	-0.0005	-0.0050	-0.0506	0.0496	0.0050	0.0005

25. Based on the table of values above, explain the end behavior of each function.

    a. $\lim\limits_{x\to\infty} f(x)$ b. $\lim\limits_{x\to-\infty} f(x)$ c. $\lim\limits_{x\to\infty} g(x)$ d. $\lim\limits_{x\to-\infty} g(x)$

    e. $\lim\limits_{x\to\infty} h(x)$ f. $\lim\limits_{x\to\infty} h(x)$

26. Consider the rule that defines the functions f, g, and h. How does the degree of the numerator compare to the degree of the denominator. (e.g., greater than, less than, or equal to.)?

27. Use the following information about the function f to answer the questions below.

    $\lim\limits_{x\to\infty} f(x) = 0$ $\lim\limits_{x\to-\infty} f(x) = 0$

    $\lim\limits_{x\to-6^+} f(x)$ DNE because $f(x)$ increases without bound

    $\lim\limits_{x\to-6^-} f(x)$ DNE because $f(x)$ decreases without bound

    $\lim\limits_{x\to6^+} f(x)$ DNE because $f(x)$ increases without bound

    $\lim\limits_{x\to6^-} f(x)$ DNE because $f(x)$ decreases without bound

    y-intercept: $(0, 5/36)$ x-intercept: $(0, 5)$

    a. Identify the vertical and horizontal asymptotes for f. State DNE if none exist.
    b. What is the domain of f
    c. Find a possible formula for f.
    d. Sketch a graph for f.

28. Use the following information about the function g to answer the questions below.

    $\lim\limits_{x\to\infty} g(x) = 7$ $\lim\limits_{x\to-\infty} g(x) = 7$

    $\lim\limits_{x\to-2^+} g(x)$ DNE because $g(x)$ decreases without bound

    $\lim\limits_{x\to-2^-} g(x)$ DNE because $g(x)$ increases without bound

    y-intercept: $(0, 7/2)$ x-intercept: $(-1, 0)$

    a. Identify the vertical and horizontal asymptotes for g. State DNE if none exist.
    b. What is the domain of g?
    c. Find a possible formula for g.
    d. Sketch a graph for g.

29. Use the following information about the function h to answer the questions below.

$\lim\limits_{x \to \infty} h(x)$ DNE because $h(x)$ increases without bound

$\lim\limits_{x \to -\infty} h(x)$ DNE because $h(x)$ decreases without bound

$\lim\limits_{x \to -\frac{9}{2}^+} h(x)$ DNE because $h(x)$ increases without bound

$\lim\limits_{x \to -\frac{9}{2}^-} h(x)$ DNE because $h(x)$ decreases without bound

y-intercept: $\left(0, -\dfrac{10}{9}\right)$ x-intercepts: $\left(-\sqrt{5}, 0\right)$ and $\left(\sqrt{5}, 0\right)$

  a. Identify the vertical and horizontal asymptotes for h. State DNE if none exist.
  b. What is the domain of h?
  c. Find a possible formula for h.
  d. Create a graph for h.

IV. CO-VARIATION OF NUMERATORS AND DENOMINATORS IN RATIONAL FUNCTIONS

30. Use the graph of f below to answer the questions that follow.

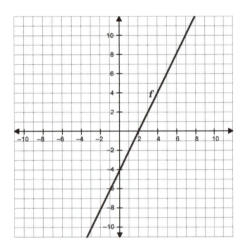

  a. Describe the important characteristics of f.

  b. Define a new function g by $g(x) = \dfrac{1}{f(x)}$.

    i.   What happens to $g(x)$ as $x \to 2^-$? As $x \to 2^+$?
    ii. What happens to $g(x)$ as x increases without bound? As x decreases without bound?
    iii. Sketch a graph of g.

  c. What, if anything, changes if the definition of g changes to become $g(x) = \dfrac{10}{f(x)}$?

      Why does this happen?

31. Use the graph of *f* below to answer the questions that follow.

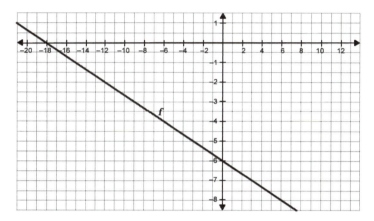

a. Describe the important characteristics of *f*.

b. Define a new function *g* by $g(x) = \dfrac{2}{f(x)}$.

    i.   What happens to $g(x)$ as $x \rightarrow -18^-$? As $x \rightarrow -18^+$?

    ii.   What happens to $g(x)$ as *x* increases without bound? As *x* decreases without bound?

    iii. Sketch a graph of *g*.

c. What, if anything, changes if the definition of *g* changes to become $g(x) = -\dfrac{2}{f(x)}$?

    Why does this happen?

32. Given the graph of *f* below, sketch the graph of *g* if $g(x) = -\dfrac{8}{f(x)}$.

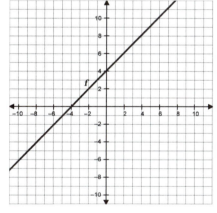

33. Given the graph of f below, sketch the graph of g if $g(x) = \dfrac{1}{f(x)}$.

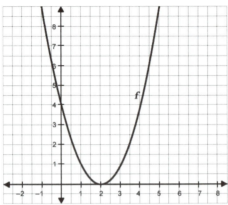

34. Given the graph of f below, sketch the graph of g if $g(x) = \dfrac{4}{f(x)}$.

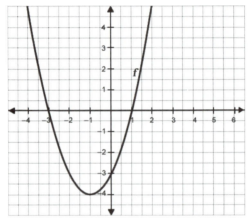

35. Given the graph of f below, sketch the graph of g if $g(x) = \dfrac{x}{f(x)}$.

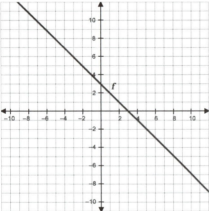

36. Given the graphs of f and g below, sketch the graph of h if $h(x) = \dfrac{g(x)}{f(x)}$.

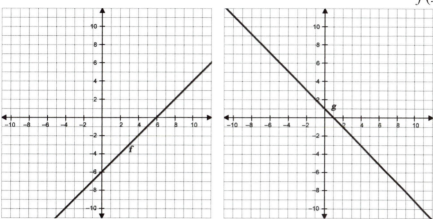

37. Given the graphs of f and g below, sketch the graph of h if $h(x) = \dfrac{g(x)}{f(x)}$.

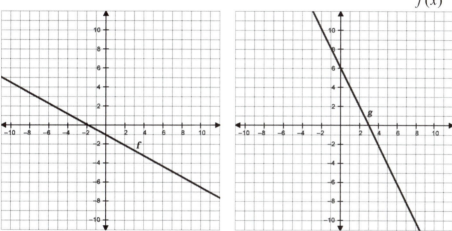

38. Given the graphs of f and g below, sketch the graph of h if $h(x) = \dfrac{f(x)}{g(x)}$.

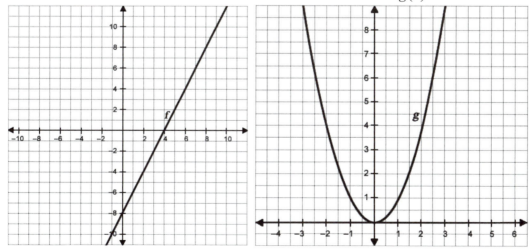

V. SYMBOLIC OPERATIONS
Note: If you need review or want to see sample worked problems like the ones below, see Module 1, Simplifying Rational Expressions, Operations on Rational Expressions, and Factoring.

Instructions for Problems 39-44: Re-write each function in simplified form (without common factors). Be sure to give the domain of each function.

39. $f(x) = \dfrac{x^2 - 5x + 6}{3x^2 - 15x + 18}$

40. $f(x) = \dfrac{3x^4 - 9x^2}{9x^3 + 27x^2}$

41. $f(x) = \dfrac{x^3 + x^2 - 2x}{-2 + x^2 + x}$

42. $f(x) = \dfrac{4y^2 - 20y - 56}{2y^2 - 22y + 56}$

43. $f(x) = \dfrac{2x^2 + 7x - 15}{2x^2 - x - 3}$

44. $f(x) = \dfrac{-x^2 + 6x - 8}{x^2 - x - 12}$

Instructions for Problems 45-50: Perform the specified arithmetic operations on the rational expressions.

45. $\dfrac{4x}{x+4} + \dfrac{x-6}{x-9}$

46. $\dfrac{5x^2}{x^3 + 1} + \dfrac{1}{x}$

47. $\dfrac{x^2 - 1}{x - 8} - \dfrac{x + 8}{7}$

48. $\dfrac{x^2}{(x+8)^4} \cdot \dfrac{(x+8)^6}{x^3(2x-1)}$

49. $\dfrac{2x^2 - x - 1}{x + 3} \cdot \dfrac{x + 3}{x - 1}$

50. $\dfrac{(4x-2)^3}{(x+3)^2} \div \dfrac{(4x-2)^5}{(x-3)^2}$

Instructions for Problems 51-56: Rewrite the following sums and differences as products by factoring the polynomials.

51. $10(x+14)-2x(x+14)$

52. $x^2(x^2-1)+13(x^2-1)$

53. $3x(x^2+5x-14)-(x^2+5x-14)$

54. $(2x^2+6)^2(x^2-2x)^3+(2x^2+6)^3(x^2-2x)^2$

55. $(x+1)^2(x+2)^3(x+3)^4+(x+1)^3(x+2)^4(x+3)^5$

56. $(2x)^{1/2}-(2x)^{5/2}$

*1. a. What is an angle?

    b. Which of the angles below has the smallest measure? Which of the angles has the largest measure? What quantity did you focus on to determine which angle has the largest measure and which has the smallest measure?

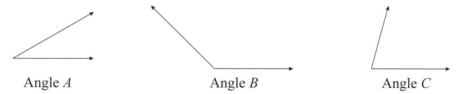

              Angle *A*   Angle *B*   Angle *C*

    c. Imagine centering circles of equal radii at the vertices of angles *A*, *B*, and *C*. Each of these three angles subtend, or cut off, a portion of the circle centered at its vertex. Refer to the following figures to determine which of the angles has the smallest measure and which has the largest measure. What quantity did you focus on to determine which angle has the largest measure and which has the smallest measure?

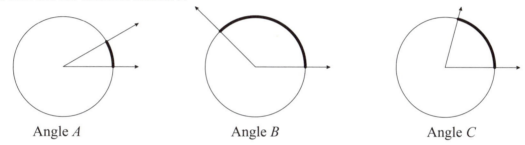

              Angle *A*   Angle *B*   Angle *C*

    d. Use your response to Part (c) to describe what it means for two angles to have the same measure?

> One way to measure an angle is to measure the length of the arc the angle subtends. Since we cannot use standard linear units such as inches or centimeters to measure an angle, Questions 2-4 explore the units we use to measure angles.

*2. a. Use a piece of string to measure each angle in units of $1/12^{th}$ of the circumference of the circle centered at its vertex. Why might it be useful to measure these angles in units of $1/12^{th}$ of the circumference of the circle centered at their respective vertices?

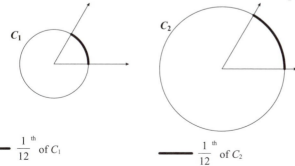

           —— $\frac{1}{12}^{th}$ of C_1 —— $\frac{1}{12}^{th}$ of C_2

b. Could you measure the two angles from Part (a) by measuring the length of the arcs these angles subtend in a standard linear unit like inches or centimeters? Explain.

3. a. Approximate the measure of the angle below in unit *m*.

b. Sketch unit *m* along the subtended arcs in the figure below so that the measure of the angle in unit *m* is maintained. What do you notice about the relationship between the length of unit *m* and the circumference of the circle on which it lies?

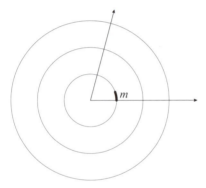

*4. Make a conjecture about the criterion that a unit of measure for the length of a subtended arc must satisfy and explain why this criterion must be satisfied.

*5. Reflect on your responses to Questions 1-4 and then describe what it means to measure an angle. (Be specific about the quantity one measures when measuring an angle and the type of units one uses to measure this quantity.)

*6. Visually estimate the measure of the angles below in the units specified.

a.

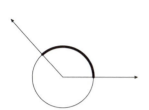

i. 1/4th of the circumference

ii. 1/6th of the circumference

iii. One whole circumference

b.

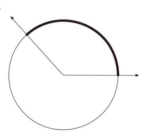

i. 1/4th of the circumference

ii. 1/6th of the circumference

iii. One whole circumference

c.

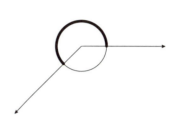

i. 1/8th of the circumference

ii. 5/8ths of the circumference

iii. One whole circumference

d. Would your answers to Parts (a)–(c) change if the radii of the circles centered at the vertex of the respective angles were twice as large? Explain.

7. Use a piece of string and a compass to complete the following tasks:
    a. Use your compass to construct a circle that has a radius equal to the length of your piece of string. (Members in your group should have strings of different lengths.)

    b. Construct an angle whose vertex is at the center of your circle and that cuts off an arc that has the length of your piece of string (one radius length). Compare the openness of your angle with that of your classmates. What do you notice?

c. Use your piece of string to create an angle that cuts off an arc that measures 2.5 radius lengths. Compare the openness of your angle with those of your classmates. What do you notice?

d. Using your piece of string, or the radius length, as a unit of measurement, how many radius lengths rotate along the circumference of your circle? Compare your answer with those of your classmates. What do you notice? Why did you get this result?

A **radian** is a unit of angle measure where one radian corresponds to an arc length equal to the radius of the corresponding circle centered at the vertex of the angle. One radian also corresponds to an arc length equal to $1/(2\pi)^{th}$ of the circumference of the circle centered at the vertex of the angle.

e. Why is a radian a convenient unit for measuring angles?

f. Does the radius of a circle centered at the vertex of an angle satisfy the criterion for a unit of angle measure discussed in Question 4? Explain.

8. Describe what it means for an angle to have the following angle measures and represent these angles on the given circle.
    a. 1 radian

    b. 2 radians

    c. 3.5 radians

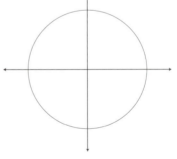

*9. Describe what it means for an angle to have a measure of
    a. 1 degree   b. 10 degrees   c. 47 degrees

A **degree** is a unit of angle measure where one degree corresponds to an arc length equal to $1/360^{th}$ of the circumference of the circle centered at the vertex of the angle.

10. a. If an angle has a measure of 1.5 radians, what percentage of the entire circle's circumference does that angle subtend?

    b. If an angle subtends an arc that is 23.87% of the entire circle's circumference, what is the measure of that angle in degrees?

    c. If an angle has a measure of 1.5 radians, what is the degree measure of that angle?

11. Convert the following angle measures given in radians to measures in degrees.
    a. 2π radians   b. 2.7 radians   c. 4 radians   d. $\pi/2$ radians

12. Define a function f to convert the measure of any angle in radians to a measure in degrees. Define your variables.

13. Convert the following angle measures given in degrees to measures in radians.
    a. 52 degrees   b. 90 degrees   c. 243 degrees

14. Define a function h to convert the measure of any angle in degrees to a measure in radians. Define your variables. How are the functions h and f (defined in Question 12) related?

1. Use a compass and a piece of string to estimate the measure of the angle below in radians. Explain what your estimate represents.

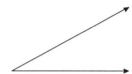

*2. a. Given that the following angle measure θ is 0.45 radians, determine the length of each arc cut off by the angle. Consider the circles to have radius lengths of 2 inches, 2.4 inches, and 2.9 inches. (*Drawing not to scale.*)

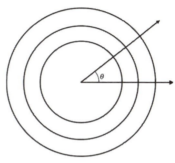

   b. Consider any circle that is centered at the vertex of an angle with a measure of θ radians. Define a formula that relates the arc length (measured in inches) cut off by the angle, the measure of the angle (measured in radians), and the radius of the circle (measured in inches). Justify your formula.

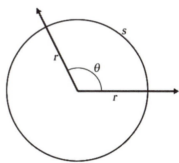

*3. What percentage of a circle's circumference is cut off by an angle that is 35 degrees plus 1.5 radians? Illustrate this angle measure below.

*4. A fellow student announced that the measure of the given angle is 1.2 inches. Is this an appropriate measure? Why or why not?

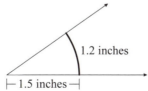

1.2 inches

├─ 1.5 inches ─┤

5. Use the given illustrations to answer the following questions (images not drawn to scale).
    a. A circle has a radius of 12 inches. An angle, whose vertex is at the circle's center, cuts the circle into two parts, with one part measuring $\frac{3\pi}{2}$ radians. How many inches is the arc length of that part of the circle?

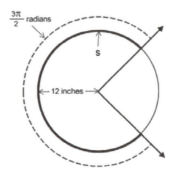

    b. A circle has a radius of r inches. An angle whose vertex is at the circle's center cuts off s inches of the circle's circumference as the terminal side of the angle opens in a counter clockwise direction from the initial side of the angle. Write a formula that conveys the relationship between the radius length r (measured in inches), the angle measure θ (measured in radians), and the arc length s (measured in inches).

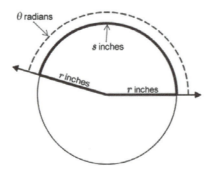

    c. A circle has a radius of 4.8 inches. An angle whose vertex is at the circle's center cuts the circle into two parts, with one part measuring 19.6 inches. What is the measure of this angle in radians?

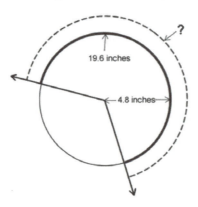

For Questions 6 and 7 imagine a bug sitting on the end of a fan blade as the blade rotates in a counter-clockwise direction. The bug is 2.6 feet from the center of the fan and is located at the 3 o'clock position as the blade begins to turn.

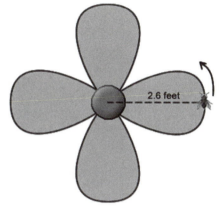

*6. a.   If the bug travels 5.2 feet around the circumference of the circle from the 3 o'clock position, what angle measure (in radians) has been swept out?

b.   If the bug travels 11.7 feet around the circumference of the circle from the 3 o'clock position, what angle measure (in radians) has been swept out?

c.   Express the angle measure θ (in radians) swept out by the fan blade in terms of the arc length traveled by the bug s and the circle's radius r (both measured in feet).

d.   If θ remains the same and the radius r of the fan doubles, how will the arc length s change?

7.   How many feet has the bug traveled around the circle swept out by the tip of the fan blade when the angle swept out by the fan blade is 0.765 radians? $\pi/2$ radians?

Imagine a bug sitting on the end of a fan blade as the blade rotates in a counter-clockwise direction. The bug is 2.6 feet from the center of the fan and is located at the 3 o'clock position as the blade begins to turn.

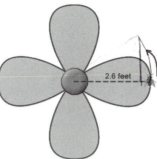

*1. What quantities might you track to represent the location of a bug as it sleeps on the tip of a rotating fan blade? List as many quantities as possible.

*2. a. Explain how the bug's vertical distance above the horizontal diameter changes as the bug travels from:
   • the 3 o'clock position to the 12 o'clock position

   • the 12 o'clock position to the 9 o'clock position

   • the 9 o'clock position to the 6 o'clock position

   • the 6 o'clock position to the 3 o'clock position

   b. As the bug travels from the 3 o'clock position to the 12 o'clock position, how does the angle measure that is swept out by the fan blade change?

   c. For successive equal changes of angle measure from the 3 o'clock position to the 12 o'clock position, how do the corresponding *changes* in vertical distance change? Identify these changes on a diagram of the situation.

$r\sin\theta$

d. Create a graph to illustrate how the bug's vertical distance above the horizontal diameter (measured in feet) co-varies with the measure of the angle θ swept out by the bug's fan blade (measured in radians). Plot points using the axes below to illustrate how the angle and the vertical distance co-vary.

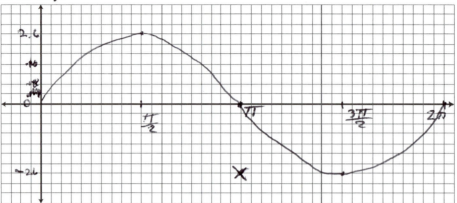

e. How will your graph in (d) change if the radius of the fan is 3.9 feet instead of 2.6 feet?

the vertical distance will increase

$\sin\theta$

f. Create a graph to illustrate how the bug's vertical distance above the horizontal diameter (measured in radius lengths) co-varies with the measure of the angle θ swept out by the bug's fan blade (measured in radians). Plot points using the axes below to illustrate how the angle and the vertical distance co-vary.

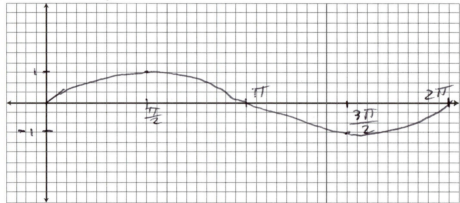

g. Explain why the graphs you created in part (d) and (f) have the same general shape.

Because it is the same function but gives different values, keeping the shape but changing the range varying.

The function you have been working with and graphing in problem 2 is called the **sine function**. It is given a special name because of how common the function is in math and the sciences. The sine function is classified as a trigonometric function.

> A function that relates an angle of rotation of a point (x, y) on the unit circle (measured from the 3'oclock position) to the distance of that point above the horizontal axis (*measured in number of radius lengths/radii*) is referred to as the **sine function.**

We see that an angle measure θ is the input to the sine function and the output $\sin(\theta)$ is a distance above the horizontal diameter measured in radius lengths. The sine function is often abbreviated by writing sin as the name of the function.

3. Label a point (x, y) on the unit circle below, then label θ and $\sin(\theta)$ as defined above.

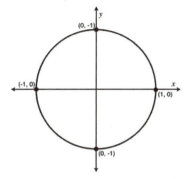

4. a. Explain how the bug's horizontal distance to the right of the vertical diameter changes as the bug travels from:
   • the 3 o'clock position to the 12 o'clock position

   • the 12 o'clock position to the 9 o'clock position

   • the 9 o'clock position to the 6 o'clock position

   • the 6 o'clock position to the 3 o'clock position

   b. For successive equal changes of angle measure from the 9 o'clock position to the 3 o'clock position, how do the corresponding *changes* in horizontal distance change? Identify these changes on a diagram of the situation.

   c. Create a graph to illustrate how the bug's *horizontal distance to the right of the vertical diameter* (measured in feet) co-varies with the measure of the angle θ swept out by the bug's fan blade (measured in radians). Plot points on the axes below to illustrate how the angle and the horizontal distance co-vary.

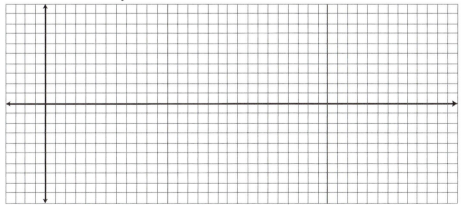

d. How will your graph in (c) change if the radius of the fan is 4.2 feet instead of 2.6 feet?

e. Create a graph to illustrate how the bug's horizontal distance to the right of the vertical diameter (measured in radius lengths) co-varies with the measure of the angle θ swept out by the bug's fan blade (measured in radians).

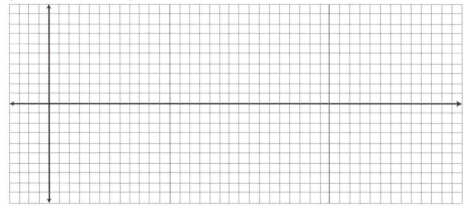

The function you've been working with in and graphing in problem 4 is called the **cosine function.**

> A function that relates an angle of rotation of a point (x, y) on the unit circle (measured from the 3'oclock position) to the distance of that point to the right of the vertical axis (*measured in number of radius lengths/radii*) is referred to as the **cosine function.**

We see that an angle measure θ is the input to the cosine function and the output $\cos(\theta)$ is a distance to the right of the vertical axis measured in radius lengths. The cosine function is often abbreviated by writing cos as the name of the function.

5. Label a point (x, y) on the unit circle below, then label θ and $\cos(\theta)$ as defined above.

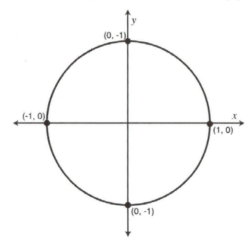

Note: Consider all θ values to be given in a number of radians.

An arctic village skiing trail has a radius of 1 kilometer. A skier started at the position (1, 0) on the coordinate axes and skied counter-clockwise.

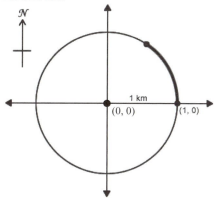

1. a. Label the points (0, 1), (–1, 0) and (0, –1) on the coordinates axes in the diagram above.

   b. What is the angle measure (in radians) of the angles that have an initial ray passing through (0, 0) and (1, 0) and a second ray that passes through (0, 0) and the point:
   i. (0, 1)   ii. (0, –1)   iii. (–1, 0)

   c. Illustrate the radian measure (in decimal form) of each of the angles (formed by the rays described above) on the coordinate axes above.

*2. An arctic village skiing trail has a radius of 1 kilometer. A skier started at the position (1, 0) on the coordinate axes and skied counter-clockwise. After skiing counter-clockwise, the skier paused for a rest at the point (0.5, 0.87).
   a. Describe what the value 0.5 represents and illustrate this measurement (value) on the axes above.

   b. Describe what the value 0.87 represents and illustrate this measurement (value) on the axes above.

3. Recall that the sine and cosine functions are defined as
    $x = \cos(\theta)$ and $y = \sin(\theta)$, given that (x, y) is a point on the
    unit circle and θ is the arc distance from (1,0) to (x,y)
    measured in radians. A unit circle with some labeled points
    is shown below.

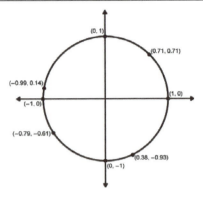

a. Use the unit circle above to generate a graph of $y = \sin(\theta)$ where θ is the angle measure formed
    by the ray through (0, 0) and (1,0) and a second ray whose origin is (0, 0) and sweeps counter-
    clockwise around the circle. (Label values on your axes and 8 points on your graph by using the
    unit circle's points above.)

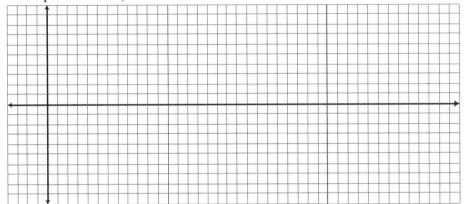

b. What is the domain of the sine function? What is the range of the sine function? As θ
    varies from π/2 to π, how does the output of the sine function vary?

c. What is the concavity of the sine function on the interval, $\frac{\pi}{2} < \theta < \pi$? Provide a rationale for your
    answer.

d. Use the unit circle above to generate a graph of $y = \cos(\theta)$.

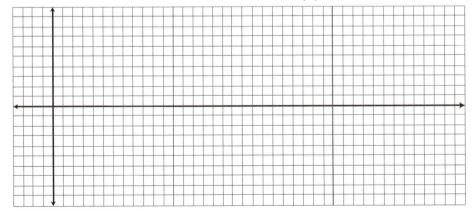

e. As θ varies from 0 to $\pi/2$, how does $y = \cos(\theta)$ vary?

*4. A second arctic village maintains a circular cross-country ski trail that has a radius of 2.5 kilometers.

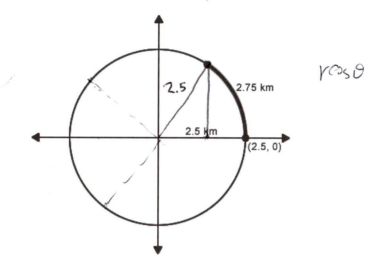

a. A skier started skiing from the position (2.5, 0) and skied counter-clockwise for 2.75 kilometers before stopping for a rest.
   i. Explain what $\frac{2.75}{2.5}$ represents in this problem.

$$\theta$$

   ii. Determine the ordered pair (measured in radii) on the coordinate axes that identifies the location where the skier rested.

$$(\cos 1.1, \sin 1.1)$$

   iii. Determine the ordered pair (measured in kilometers) on the coordinate axes that identifies the location where the skier rested.

$$\left(\frac{\cos \frac{2.75}{2.5}}{1.1}, \frac{\sin \frac{2.75}{2.5}}{1.1}\right)$$

b. The skier then continued along the trail, traveling counter-clockwise until he had traveled for a *total* of 10 kilometers before stopping for another rest.
   i. Determine the ordered pair (measured in radii) on the coordinate axes that identifies this location where the skier rested.

$$(\cos 5.1, \sin 5.1)$$

$$\frac{12.75}{2.5}$$
$$\theta = 5.1$$

   ii. Determine the ordered pair (measured in kilometers) on the coordinate axes that identifies this location where the skier rested.

$$(\cos 2.5, \sin 2.5)$$

c. Suppose that a second skier started skiing from the position (2.5, 0) and skied *clockwise* for 5 kilometers before stopping for a rest.

   i. Determine the ordered pair (measured in radii) on the coordinate axes that identifies this location where the skier rested.

$$(\cos 2, \ -\sin 2)$$

   ii. Determine the ordered pair (measured in kilometers) on the coordinate axes that identifies this location where the skier rested.

$$\left(2.5 \cos \tfrac{2}{2.5}, \ 2.5 \cdot \sin 2\right)$$

5. Consider a coordinate system where the coordinate values of a point are measured as a number of radii. What is the general form of an ordered pair (expressed in a number of radii) of any point on a circle that has a radius length of *r* kilometers and an angle measure of θ radians?

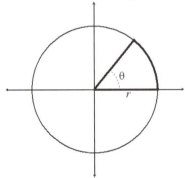

*6. a. Construct angles on the unit circle with these radian measures: $\frac{\pi}{6}, \frac{5\pi}{6}, \frac{7\pi}{6}, \frac{11\pi}{6}$.

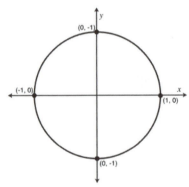

b. Using your calculator, determine the coordinates corresponding to the angle with a measure of $\frac{\pi}{6}$ radians. Describe what these coordinates represent.

c. Using the coordinates determined in part (b) above (not your calculator), determine the coordinates that correspond to $\frac{5\pi}{6}, \frac{7\pi}{6}, \frac{11\pi}{6}$ radians (angle measures). Explain how you determined these coordinates.

José extends his arm straight out, holding a 2.3-foot string with a ball on the end. He twirls the ball around in a circle with his hand at the center, so that the plane in which it is twirling is perpendicular to the ground. Answer the following questions assuming the ball twirls counter-clockwise starting at the 3 o'clock position.

1. Label quantities on the diagram that could be used to describe the position of the ball as it rotates around José's hand.

*2. Define the algebraic form of the function *f* that relates the ball's *vertical distance above* Jose's hand (in feet) as a function of the measure of the angle swept out by the ball and string (measured in radians) as the ball twirls counter-clockwise from a 3 o'clock position. (Define the relevant variables.)

*3. Suppose the ball that Jose was twirling travels one radian per second.
    a. How many radians θ does the ball and string sweep out in *t* seconds?

    b. Define a function *g* that relates the ball's *vertical distance above* José's hand as a function of the number of seconds elapsed. (Define the relevant variables.)

    c. Over what time interval does the ball complete one revolution?

*4. Suppose the ball travels 2 radians per second.
    a. How many radians θ does the ball and string sweep out in *t* seconds?

    b. Define a function *h* that relates the ball's *vertical distance above* Jose's hand as a function of the number of seconds elapsed. (Define the relevant variables.)

    c. Over what time interval does the ball complete one revolution?

*5. Suppose the ball travels 0.5 radians per second.
    a. How many radians θ does the ball and string sweep out in *t* seconds?

b. Define a function *j* that relates the ball's *vertical distance above* José's hand as a function of the number of seconds elapsed. (Define the relevant variables.)

c. Over what time interval does the ball complete one revolution?

*6. Graph the functions determined in questions 3-5 on the axes below. On each graph illustrate the interval of input values on which the function values complete one full cycle (i.e., the interval of *t* for which the ball completes one revolution).

a. Complete the following table of values and then graph *g* on the given axes (see question 3). Label your axes.

g(t) =	
t	*g(t)*
0	
π/2	
π	
3 π/2	
2 π	
5 π/2	
3 π	
7 π/2	
4 π	

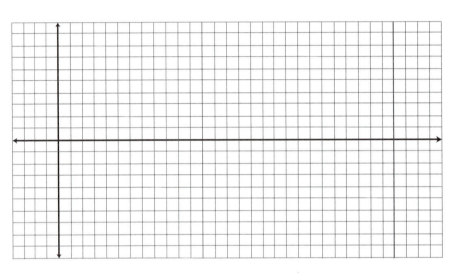

b. Complete the following table of values and then graph *h* on the given axes (see question 4). Label your axes.

h(t) =	
t	*h(t)*
0	
π/4	
π/2	
3 π/4	
π	
5 π/4	
3 π/2	
7 π/4	
2π	

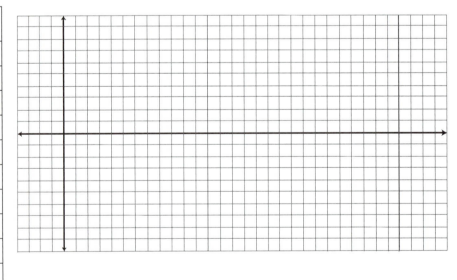

c. Complete the following table of values and then graph *j* on the given axes (See question 5). Label your axes.

$j(t) =$	
t	*j(t)*
0	
π/2	
π	
3 π/2	
2 π	
5 π/2	
3 π	
7 π/2	
4 π	

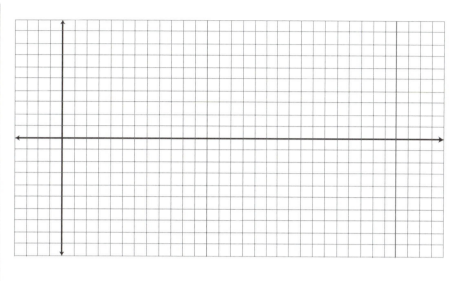

The **period** of a trigonometric function is the interval of input needed to complete a full cycle of output values.

---

d. Describe the reasoning you used to determine the period of the functions in (a), (b), and (c) above.

7. Suppose the ball and string sweeps out 3.8 radians per second.
   a. How many radians *θ* does the ball and string sweep out in *t* seconds?

   b. Define a function *k* that relates the ball's *vertical distance above* José's hand as a function of the number of seconds elapsed. (Define the relevant variables.)

   c. Over what time interval does the ball complete one revolution?

8. Suppose you determine that it takes the ball 20 seconds to complete one revolution.
   a. Determine the speed of the ball in radians/second.

   b. Define a function *m* that relates the ball's *vertical distance above* José's hand as a function of the number of seconds elapsed. (Define the relevant variables.)

---

*1. A Ferris wheel has a radius of 52 feet and the horizontal diameter is located 58 feet above the ground.

    a.   If the Ferris wheel rotates at a constant rate of ¼ radian per minute, how many minutes does it take for the Ferris wheel to make one full revolution?

    b.   Define a function g to represent the distance (in feet) of a Ferris wheel bucket above the *ground* as a function of the number of seconds, t, since the Ferris wheel began rotating from the 3 o'clock position. Assume the Ferris wheel rotates at a constant rate of ¼ radian per minute.

    c.   Suppose the Ferris wheel rotates at a constant rate of 3 radians per minute. How many minutes will it take for the Ferris wheel to make one full revolution?

    d.   What might the function $h(t) = \sin(2t)$ represent in this situation? (Hint: Consider the meaning of the value of $2t$.)

2.   Why is the sine function useful? What two quantities does the sine function relate?

3.   What is the period of the function f, defined by $f(\theta) = \sin(\theta)$? What does the period of f represent?

4.   A bike is on the rack at a bike shop so that the technician can turn the pedal with his hands, causing the bike's back wheel to rotate as he turns the pedal. For every 1 radian that the technician rotates the pedal counterclockwise (CCW), the wheel spins 4 radians CCW.

    a. Make an illustration of the circular turning path of the pedal and the circular turning path of the back tire to depict their relative sizes.

    b. Suppose the pedal and a tire valve on the wheel both start at the 3 o'clock positions on their respective circle of motion. The tire valve is 12.5 inches from the center of the wheel's rotation. Let θ = the angle that the *pedal* sweeps counterclockwise (CCW) from the 3 o'clock position. Illustrate the location of the pedal and the tire valve on the circles you created in part (a).

    c. Write a function f that outputs the *horizontal distance of the pedal from the vertical diameter* in radius lengths as the pedal rotate θ radian.

d. Write an expression to define the *angle rotation of the tire valve* in terms of the *angle of rotation of the pedal θ*.

e. Use your expression in part (d) to write a function *g* that outputs the *horizontal distance of the tire valve from the wheel's vertical diameter* in radius lengths if the input quantity is θ.

f. What interval of θ will cause the <u>pedal</u> to make a full rotation? What interval of θ will cause the <u>tire valve</u> to make a full rotation? Explain your reasoning.

g. Graph *f* and *g* on the same set of axes below.

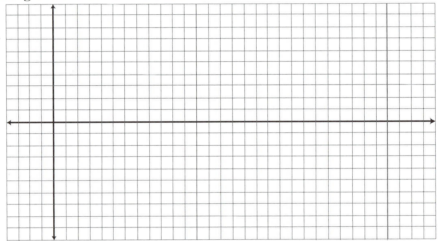

h. What is the period of *f*? What is the period of *g*? Explain what the period of *g* means in the context of the pedal and the tire valve.

i. The tire valve always sweeps an angle 4 times as large as the pedal. Explain why the period of *g* is *less* than the period of *f*.

5. Suppose the bike technician changes gears on the same bike described in question (4), so that the wheel has a different rate of rotation relative to the pedal rotation. Consider the function h defined by $h(\theta) = 12.5\cos\left(\frac{\theta}{2}\right)$ that defines the horizontal distance of the *tire valve* on the wheel from the vertical axes (in inches) in terms of the *pedal angle* θ (in radians) swept out counter clockwise from the 3 o'clock position.

   a. What does the quantity $\frac{\theta}{2}$ represent in the above definition of h?

   b. What interval of the quantity $\frac{\theta}{2}$ results in one full cycle of output values for the function h? Explain your reasoning.

   c. Use your answer in part (b) to find the interval of θ that results in one full cycle of output values of $h(\theta)$.

   d. Which of the answers above (b or c) is the period of h? Explain the meaning of the period in the context of the rotating pedal and tire valve.

   e. Describe what the output quantity of $h(\theta)$ represents.

*6. The function h is defined by $h(\theta) = \sin(4\theta) + 2$ where θ is an angle measure (in radians).

   a. Complete the following table.

θ	$h(\theta)$
0	
$\pi/8$	
$\pi/4$	
$3\pi/8$	
$\pi/2$	
$5\pi/8$	
$3\pi/4$	

   b. Graph h on the axes below. (Label your axes.)

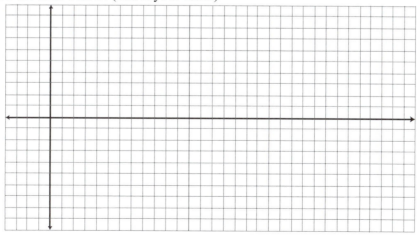

c.   What is the period of *h*? What does the period of *h* represent?

d.   What is the amplitude of *h*?

7.   a.   Let $j(\theta) = \sin\left(\frac{\theta}{2}\right)$. Since the argument to the sine function ($\frac{\theta}{2}$ in this case) always creates a full cycle of output values every 2π radians, how does θ vary as the argument to the sine function, $\frac{\theta}{2}$, varies by 2π?

b.   Given the functions *f* and *j* are defined as $f(\theta) = \sin(\theta)$ and $j(\theta) = \sin\left(\frac{\theta}{2}\right)$, construct graphs of *f* and *j* on the same axes. (Label the axes before constructing your graphs.)

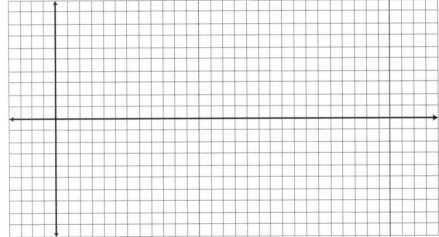

c.   Explain how you determined the period of *j*. Identify the period of *j* on the graph in part (b) by identifying two intervals of input values that correspond to the period.

*8.   Let $k(\theta) = 2\sin\left(\frac{3}{7}\theta\right) + 5$. What is the period of *k*? Explain.

*9. Consider the graph of the function *h* below:

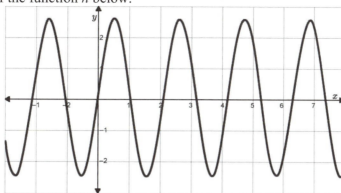

a. What is the approximate value of the period of *h*?

b. What is the approximate value of the amplitude of *h*?

c. Define a formula for the function *h*.

10. A picture of a water wheel (used for power in the 18th century) is illustrated below.

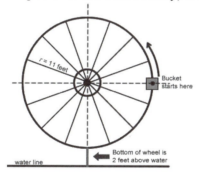

a. Sketch a graph of the *height* of the bucket above the water (in feet) as a function of the angle of rotation (measured in radians) of the bucket from the 3 o'clock position. (Define your variables and label your axes.)

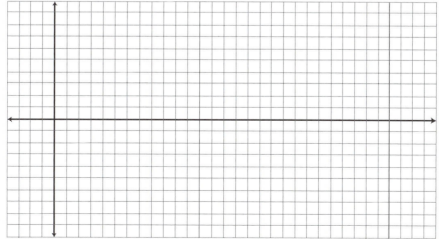

*b. Define a function *f* that represents the bucket's distance *d* above the water (in feet) as a function of the angle of rotation (measured in radians) of the bucket from the 3 o'clock position. (Define your variables.)

*c. Define a function *g* that represents the bucket's distance *d* above the water (in feet) as a function of the *number of feet* the bucket has rotated counter-clockwise from the 3 o'clock position. (Define your variables.)

1. Consider the graph of the function *f* below:

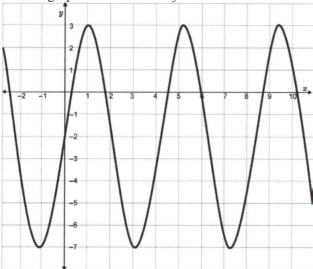

a. What is the approximate value of the period of *f*?

b. What is the approximate value of the amplitude of *f*?

c. Define a formula for the function *f*.

2. A bike is on the rack at a bike shop so that the tech can turn the pedal with his hands, causing the bike's back wheel to rotate as well. On this bike the circular path swept out by the pedal is exactly the same size as the circular path swept out by the tire. Suppose that when the pedal starts at the 3 o'clock position, the tire valve on the wheel is $\frac{\pi}{4}$ radians clockwise from 3 o'clock (in quadrant IV).

a. Assume that on this bike every 1 radian rotation of the pedal counter clockwise (CCW), the back wheel concurrently rotates 1 radian CCW.

Let θ = the angle measure of the *pedal* CCW from the 3 o'clock position, in radians.

Fill in the following table to show the angle measure of the tire valve from the 3 o'clock position in radians corresponding to various values of θ. Use *exact* values for all angle measures.

When the pedal's angle measure θ is…	…the tire valve's angle measure is…
0	$-\pi/4$
$\pi/6$	
$\pi/2$	
2	
$4\pi/3$	
5.5	
2π	

b. What is the angle measure of the tire valve from the 3 o'clock position when the pedal angle measure is θ radians from 3 o'clock? Write an expression.

c. As the pedal and wheel turn together, is the tire valve 'ahead' or 'behind' the pedal? Use your expression in part b) to explain your answer.

d. Define a function *P* that outputs the vertical distance above the horizontal diameter of the pedal (in radii), in terms of the pedal angle measure *θ*.

e. Define a function *T* that outputs the vertical distance above the horizontal diameter of the tire valve (in radii), in terms of the **pedal** angle measure *θ*.

f. Fill in the missing table entries with *exact* values for the output values of *P* and *T*.

θ	$P(\theta)$	$T(\theta)$
0		
π/4		
π/2		
3π/4		
π		
5π/4		
3π/2		
7π/4		
2π		
9π/4		

g. Sketch and label the graphs of the functions *P* and *T* on the axes below. Sketch only the cycle of each function that corresponds to the values in the table.

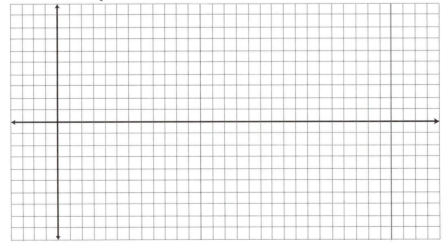

h. Put your finger at (0, 0) on your graph and run it slowly along the input axis to represent increasing θ values. Does the cycle of *T* begin and transpire before or after the cycle of *P*? Does this agree with your answer to part c)? Explain.

i. The graph of *T* represents a shift to the right of the graph of *P* by $\pi/4$. Why does it make sense for *T* to be thought of as shifted to the right? How does the function rule for *T* represent a right-shift of the graph?

3. For the same bicycle context as in #2, suppose that when the pedal is in the 3 o'clock position, the tire valve on the wheel is $\pi/6$ radians counter-clockwise from 3 o'clock (in quadrant I). The tire still rotates one radian CCW for every one radian the pedal rotates CCW, and θ = the angle measure of the *pedal* CCW from the 3 o'clock position, in radians.
   a. Consider the function $f(\theta) = \cos(\theta)$. Explain the meaning of this function in the bicycle context.

   b. Sketch *f* on the axes below if the domain of *f* is $0 < \theta < 2\pi$

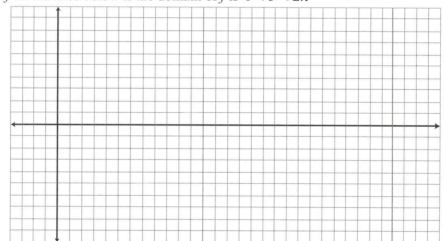

   c. As the pedal and wheel rotate together, where is the tire valve on its circle of rotation compared to where the pedal is on its circle of rotation? Make your comparison in terms of angle measure(s).

   d. Write an expression that gives the CCW angle of the tire valve from the 3 o'clock position in terms of θ, the CCW angle of the pedal from the 3 o-clock position. Explain your answer.

   e. Write a function *g* that outputs the tire valve's horizontal distance from the vertical diameter in radius lengths as a function of θ.

f. On your graph in part (b), the cycle of f (for the pedal) begins at $\theta = 0$. Given that the tire valve is $\frac{\pi}{6}$ *ahead* of the pedal, will the corresponding cycle of g (for the tire valve) start *before* or *after* this cycle of f on the graph? At what value of θ will this cycle of g start? Explain. (Hint: Put your finger at $\left(\frac{-\pi}{2}, 0\right)$ on your graph and run it slowly along the input axis to represent increasing θ values. If the tire valve is $\frac{\pi}{6}$ *ahead*, where will it begin its cycle relative to the pedal?)

g. On your graph in part (b), sketch this cycle of g representing the tire valve making one full cycle, from the 3 o'clock position, to the 3 o'clock position.

h. Is the graph of g the same as the graph of f shifted left, or shifted right? Explain your answer.

4. A bike is on the rack at a bike shop so that if the tech turns the pedal counter clockwise, the bike's back wheel turns CCW also. The tire valve on the back wheel is 12.5 inches from the center of the wheel's rotation. Let θ = the angle measure of the *pedal* CCW from the 3 o'clock position, in radians.

For each of the following pairs of functions, the second function is a transformation of the first. *The second function of each pair* has an output quantity related to the *tire valve* on the back wheel of the bicycle.

State the meaning of each function and describe how each pair of functions and their graphs are related.

a. $f(\theta) = \sin(\theta)$ $\qquad\qquad g(\theta) = 12.5\sin(\theta)$

b. $h(\theta) = 12.5\cos(\theta)$ $\qquad\qquad j(\theta) = 12.5\cos\left(\theta - \frac{\pi}{3}\right)$

c. $p(\theta) = 12.5\sin(\theta)$ $\qquad\qquad q(\theta) = 12.5\sin(2\theta)$

d. $r(\theta) = \cos(\theta)$ $\qquad\qquad s(\theta) = \cos\left(\frac{\theta}{3}\right)$

e. $v(\theta) = \sin(\theta)$ $\qquad\qquad w(\theta) = \sin\left(\theta + \frac{\pi}{2}\right)$

1. Consider the graph below, which shows the first quadrant of the coordinate plane [the bottom left corner is (0, 0) A = (1,0) and J=(0,1)]. Assume the distance between consecutive points on the curve represents 1/9th of a quarter circle. The radius of the circle is one unit.

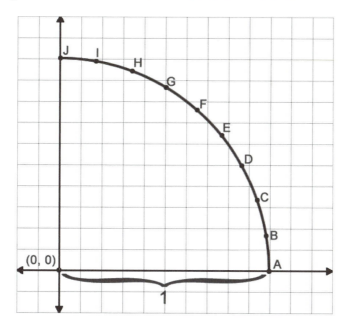

*a. Consider the angle that is formed by the initial ray (passing through the origin and point A) and the terminal ray (passing through origin and point B).
   i. What is the measure of this angle in radians?

$$\frac{1}{36}(2\pi) \qquad \frac{\pi}{18}$$

   ii. What is the ordered pair (x,y) describing the location of point B?

$$\left(\cos\frac{\pi}{18}, \ \sin\frac{\pi}{18}\right)$$

   iii. What is the slope of the terminal ray (the ray passing through the origin and point B?

$$\frac{\sin\frac{\pi}{18}}{\cos\frac{\pi}{18}}$$

b. Consider the angle that is formed by the initial ray (passing through the origin and point A) and the terminal ray (passing through origin and point C).
   i. What is the measure of this angle in radians?

   ii. What is the ordered pair (x,y) describing the location of point C?

   iii. What is the slope of the terminal ray (the ray passing through the origin and point C)?

*c. Complete the table below.

Point	Measure of the angle formed by the initial ray and the terminal ray passing through the given point (in radians), θ	Ordered pair describing the point's location (x, y)	Slope of the terminal ray passing through the given point, m
A	0	(1,0)	About 0
B	π/18	About (0.985, 0.174)	About 0.177
C	2 π/18 = π/9	About (0.940, 0.342)	About 0.364
D	3π/18	About (0.866, 0.5)	About 0.577
E	4π/18 = 2 π/9	About (0.766, 0.643)	About 0.839
F	5 π/18	About (0.643 0.766)	About 1.192
G	6 π/18 = π/3	About (.5, 0.866)	About 1.732
H	7 π/18	About (0.342, 0.940)	About 2.747
I	8π/18 = 4 π/9	About (0.174, 0.985)	About 5.661
J	9 π/18 = π/2	(0,1)	undefined

*d. What happens to the slope of the terminal ray as the value of θ approaches $\frac{\pi}{2}$?

It exponentely increases.

e. For what angle measure θ (Between $\theta = \pi$ and $\theta = \frac{\pi}{2}$) will the slope of the terminal ray be 1?

2. Now let's think about how the slope of the terminal ray and the angle measure co-vary in the second quadrant $\left(\text{that is, when } \frac{\pi}{2} \le \theta \le \pi\right)$.

*a. When $\frac{\pi}{2} < \theta < \pi$, is the slope of the terminal ray positive or negative? Justify your answer.

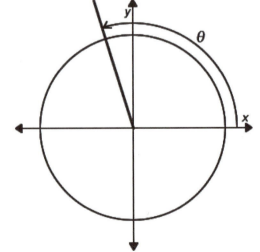

*b. As θ increases from $\theta = \frac{\pi}{2}$ to $\theta = \pi$, how does the slope of the terminal ray vary?

c. For what angle measure θ (Between $\theta = \frac{\pi}{2}$ and $\theta = \pi$) will the slope of the terminal ray be –1?

*3. Now let's think about how the slope of the terminal ray and the angle measure co-vary in the third quadrant.
   a.   When $\pi < \theta < \frac{3\pi}{2}$, is the slope of the terminal ray positive or negative? Justify your answer.

   b.   As θ increases from $\theta = \pi$ to $\theta = \frac{3\pi}{2}$, how does the slope of the terminal ray vary?

*4. Now let's think about how the slope of the terminal ray and the angle measure co-vary in the fourth quadrant.
   a.   When $\frac{3\pi}{2} < \theta < 2\pi$, is the slope of the terminal ray positive or negative? Justify your answer.

   b.   As θ increases from $\theta = \frac{3\pi}{2}$ to $\theta = 2\pi$, how does the slope of the terminal ray vary?

5. For what angle measure(s) θ (between $\theta = \pi$ and $\theta = 2\pi$) will the slope of the terminal ray be
   a.   1?

   b. -1?

   c.   greater than 1?

   d.   less than -1?

The function you've been working with is called *the* **tangent function** and is abbreviated as $f(\theta) = \tan(\theta)$. The tangent function is a function that takes as its input an angle measure (from the 3 o'clock position) and outputs the slope of the terminal ray of the angle.

---

*6. Without using a calculator determine if each of the following statements is true or false. Justify your answer by drawing a diagram or with a written explanation.

   a. $\tan\left(\dfrac{2\pi}{3}\right) < 0$ True

   b. $\tan\left(\dfrac{\pi}{5}\right) < 0$ False

   c. $0 < \tan\left(\dfrac{4\pi}{3}\right)$ True

   d. $\tan\left(\dfrac{7\pi}{8}\right) > 0$ False

7. Use a calculator to evaluate each of the following and then interpret the meaning of your answer.

   a. $\tan(75^\circ) =$

   b. $\tan\left(\frac{\pi}{7}\right) =$

   c. $\tan\left(\frac{3\pi}{5}\right) =$

8. Sketch a graph of the slope of the terminal ray versus the angle measure as the angle measure varies from 0 radians to 2π radians.

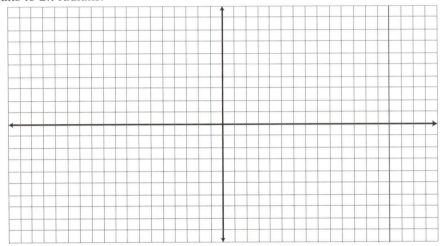

9. The tangent function is periodic because it repeats at regular intervals.

   a. What is the period of the tangent function?

   b. Where are the vertical asymptotes located and why do they occur for these values of θ?

   c. Use the following diagrams to help you explain why the slopes of the terminal ray are the same whenever the angle measure differs by π radians (or 180°).

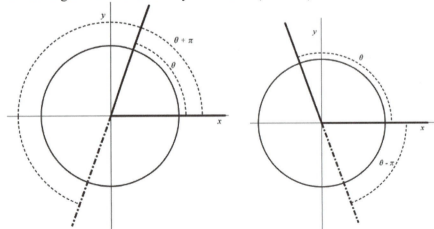

*10. In Exercise #1 you were asked to calculate the slope of the terminal ray of the angle when you knew that the ray began at the origin and passed through a point on the circle. For example, if the terminal ray passed through the point (0.1736, 0.9848) then the slope was

$$m = \frac{0.9848 - 0}{0.1736 - 0} = \frac{0.9848}{0.1736} \approx 5.673$$

Suppose you know that an angle measures θ radians using a circle of radius 5.

a. How can you find the point (x, y) on the circle that the terminal ray passes through?

b. Using your answer to part (a), complete the following statement by writing $\tan(\theta)$ in terms of $\cos(\theta)$ and $\sin(\theta)$. (In other words, when we know the value of $\cos(\theta)$ and $\sin(\theta)$, how can we determine the value of $\tan(\theta)$?)

$\tan(\theta) =$

*11. In earlier investigations we determined that we can find a point (x, y) on the circle using $x = r \cdot \cos(\theta)$ and $y = r \cdot \sin(\theta)$ (instead of just $x = r \cdot \cos(\theta)$ and $y = r \cdot \sin(\theta)$) where r is the radius of the circle. If the radius of a circle is r, is the slope of the terminal side $m = r \cdot \tan(\theta)$ or $m = \tan(\theta)$. Justify your answer.

*1. a. Complete the function machine diagrams below.

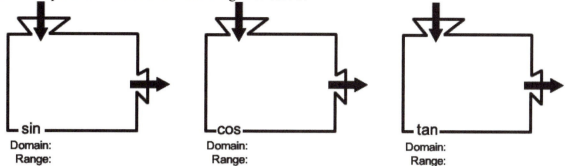

b. What is the input quantity for the inverse relation of the sine function?

c. What is the output quantity for the inverse relation of the sine function?

d. Is the inverse relation of the sine function a function (i.e., does each vertical distance above the horizontal diameter produce a unique angle measure)? Explain.

When considering the **inverse sine function** we first restrict the domain of the sine function to $\frac{-\pi}{2} \le \theta \le \frac{\pi}{2}$ and then consider the inverse. The inverse sine function takes as its input the vertical distance, from where the terminal ray (of the angle) intersects the unit circle, above the horizontal diameter (in radius lengths) and outputs an angle measure. We call this inverse the arcsine or inverse sine function and write $\arcsin(y) = \theta$ or $\sin^{-1}(y) = \theta$.

When considering the **inverse cosine function, also called the arccosine function** we first restrict the domain of the cosine function to $0 \le \theta \le \pi$. The inverse cosine function takes as its input the horizontal distance, from where the terminal ray (of the angle) intersects the unit circle, to the right of the vertical diameter (in radius lengths) and outputs an angle measure. We call this inverse the arccosine or inverse cosine function and write $\arccos(x) = \theta$ or $\cos^{-1}(x) = \theta$.

When considering the **inverse of the tangent function** we first restrict the domain of the tangent function to $\frac{-\pi}{2} < \theta < \frac{\pi}{2}$ and then consider the inverse. The inverse tangent function takes as its input the slope of the terminal ray (of the angle) and outputs an angle measure. We call this inverse the arctangent or inverse tangent function and write $\arctan(m) = \theta$ or $\tan^{-1}(m) = \theta$.

---

e. Complete the following function machine diagrams below.

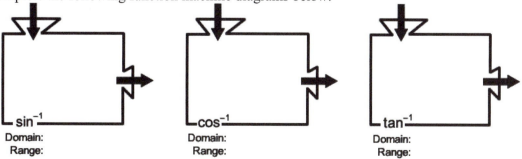

© 2015 Carlson, Oehrtman, and Moore

*2. Demonstrate on the following circles how one could estimate the value of $\sin^{-1}(0.25)$. Then use your calculator to evaluate $\sin^{-1}(0.25)$. Explain why the answer your calculator gives makes sense.

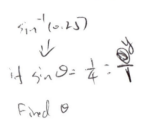

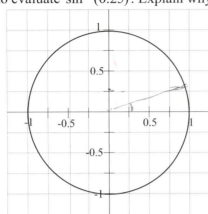

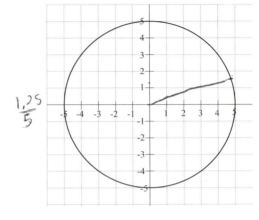

$\frac{1.25}{5}$

*3. Using the following circles estimate the value of $\cos^{-1}(-0.75)$. Explain your approach.
Then use your calculator to evaluate $\cos^{-1}(-0.75)$. Explain why the answer displayed by your calculator makes sense.

$\frac{-.75}{1}$

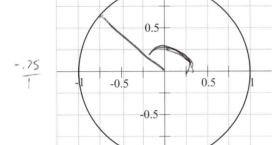

$\frac{-3.75}{5}$

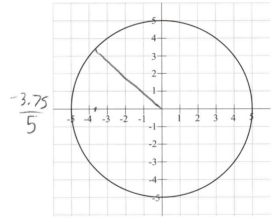

*4. On the following circles how one could estimate the value of $\tan^{-1}(2)$. Then use your calculator to evaluate $\tan^{-1}(2)$. Explain why the answer your calculator gives makes sense.

$\frac{x}{y} \quad \frac{1}{.5}$

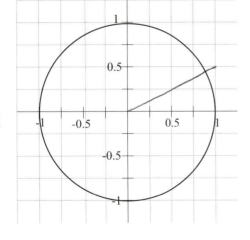

$\frac{5}{2.5}$

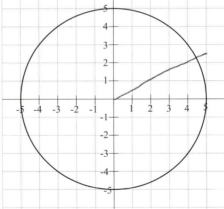

*5. Savanna is sitting in a bucket of a Ferris wheel at the 3 o'clock position. The Ferris wheel has a radius of 62 feet. The Ferris wheel begins moving counterclockwise.
   a. What angle(s) has Savanna rotated from the 3 o'clock position when she is 42 feet above the horizontal diameter of the Ferris wheel?

   b. What angle(s) has Savanna rotated when she is −20 feet to the right of the vertical diameter of the Ferris wheel?

6. Determine the measure of the angle θ (in radians) indicated on the circle.

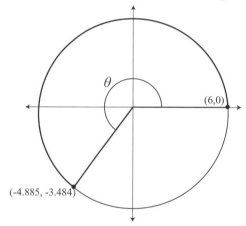

*7. A skier skied on a circular trail, starting at the position (2.5, 0) on the circle, and ended at the position (−2.3775, 0.7725). How many kilometers did the skier ski? Calculate the distance skated by the skier by determining values for both the inverse sine and inverse cosine functions. Explain your approach and why it is valid.

8. Given the following point on a circle of radius 4 meters centered at the origin, determine the corresponding counter-clockwise arc length (in the same units as the coordinates of each point) from the 3 o'clock position.
   a. (1.874, −3.534) meters

   b. (−0.837, −0.547) radii

*9. Determine *values* of θ, $0 \le \theta \le 2\pi$, that make the following statements true. If there are no such values of θ, say why not.

a. $3\cos(\theta) = 1$

b. $\dfrac{1}{4}\cos(\theta) = -0.2$

c. $5\cos(\theta) = -5$

d. $2\sin(\theta) = 1$

e. $\sin(\theta) = .8$

f. $\frac{1}{3}\sin(\theta) = 0.5$

g. $\tan(\theta) = 1$

h. $3\tan(2\theta) = -8$

i. $2\tan(\theta) = -\dfrac{1}{3}$

*10. Given the following circle and undetermined angle measures of α and θ radians, answer the following.

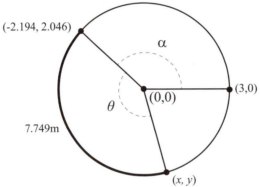

a. What is the value of θ, the measure of the angle indicated in the figure above?

b. What quantity does the value $\frac{-2.194}{3}$ represent? What is the unit of measure for this quantity?

c. What quantity does the value $\frac{2.046}{3}$ represent? What is the unit of measure for this quantity?

d. What is the value of α, the measure of the angle indicated in the figure above?

e. Determine the values for x and y indicated in the figure above.

I. ANGLE MEASURE

1. Describe the meaning of an angle measuring θ. Construct an angle and illustrate the angle's measure θ on your diagram.

2. A pizza chef wants to design a pizza cutter that will cut his 15-inch diameter circular pizzas into 11 equal pieces. He has no access to a protractor. How might he design his pizza cutter?

3. Explain, in terms of arc length and circumference, what it means for an angle to have a measure of:
   a. 45 degrees   b. 191.4 degrees

4. The "grad" is a unit of angle measure that is sometimes used in France. One grad is the measure of an angle that cuts off an arc that is $\frac{1}{400}$th of a circle's circumference. An angle that rotates a circle's circumference is 400 grads. (Recall that one degree is an angle with vertex at the origin that cuts off an arc that is $\frac{1}{360}$th of a circle's circumference and every circle's circumference is 360 degrees.)
   a. Using measures of circumference and arc length, explain how to make a protractor that measures an angle's openness in grads.
   b. If the protractor has a radius of 4.3 inches, determine the arc length in inches of:
      i. 1 grad   ii. 10 grads   iii. 200 grads   iv. 400 grads
   c. An angle measuring 50 grads subtends what percent (or fractional part) of a circle's circumference? Explain your answer.
   d. How many degrees are equivalent to 50 grads? 100 grads? 10.2 grads?
   e. Define a function that converts a number of grads to a number of degrees. Explain how you determined this function.

5. Name and define your own unit of angle measurement (e.g., describe how many of these units mark off any circle's circumference, making it at least 10 units). Describe how to create a protractor to measure angles using the unit you defined.

6. Is there a benefit to our choice of units for measuring angles? For example, is there a benefit to using 360 degrees or 400 grads to rotate a circle compared to some unit that cuts off an arc that is $\frac{1}{13}$th or $\frac{1}{70}$th of a circle's circumference?

7. Explain in terms of arc length and radius what it means for an angle to have a measure of:
   a. 1 radian   b. 2.1 radians   c. 3 radians   d. π radians   e. $\frac{1}{2}\pi$ radians

8. Draw a circle. Now form a central angle (the angle's vertex is at the center of the circle) that has a measure of 1 radian. Describe your approach.

9. a. Convert the following angle measures from degrees to radians. Explain your method and why it works.
      i. $37°$ ii. $310°$ iii. $715°$ iv. $90°$
   b. Convert the following angle measures from radians to degrees. Explain your method and why it works.
      i. 3.7 radians   ii. 0.5π radians   iii. 6.28 radians   iv. 2π radians   v. 5.5 radians
   c. Define a function f that accepts an angle measure in degrees as input and returns a unique angle measure in radians as output. Define your variables.

(question continues on next page)

d. Define a function g that accepts an angle measure in radians as input and returns a unique angle measure in degrees as output. Define your variables.

e. Describe the process of composing *f* and *g*. Evaluate *g*(*f*(*x*)).

f. If the openness of an angle varies from 35° to 112°, the openness of the angle changes by how many radians? Draw a circle and illustrate the change in angle measure.

10. The *barc* is a unit of angle measure. One barc is the measure of an angle that cuts off an arc that is $\frac{1}{240}$ th of a circle's circumference. An angle that rotates a circle's circumference is 240 barcs.

a. Using measures of circumference and arc length, explain how to make a protractor that measure an angle's openness in barcs.

b. If a protractor for measuring angles in barcs has a radius of 7.1 inches, explain how a circle can be used to create this protractor. What is the arc length in inches on this protractor of 1 barc? 50 barcs? 100 barcs?

c. How many radians are equivalent to 200 barcs? 521.2 barcs?

d. Define a function that converts a number of degrees to a number of barcs. Define a function that converts a number of barcs to a number of radians.

e. Using the two functions above, determine a function that converts a number of degrees to a number of radians.

II. ANGLE MEASURE IN CONTEXT

11. a. Determine the percentage of a circle's circumference cut off by an angle that has a measure of 22.3 degrees plus 301.1 grads (recall that any circle's circumference is 360 degrees or 400 grads).

b. The measure of the angle is: i) how many degrees? ii) how many radians?

12. Three angles have their vertex at the center of a circle with a circumference of 21 inches. Their measures are given below. Determine the linear measure, in inches, of the arc cut off by each angle. Make a sketch to illustrate each measure.

a. 192 degrees

b. 4.7 radians

c. $\frac{\pi}{6}$ radians

13. a. Determine the measure of an angle (in radians) that cuts off an arc length of 5.1 inches in a circle with a 5.1 inch radius.

b. Determine the measure of an angle (in radians) that cuts off an arc length of 10.2 inches in a circle with a 5.1 inch radius.

c. Determine the measure of an angle (in radians) that cuts off an arc length of 21.3 inches in a circle with a 5.1 inch radius.

14. Answer the following questions.

a. Determine the value of *s*, the arc length (measured in inches) cut off by the angle with a measure of 2.1 radians.

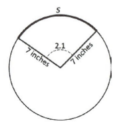

(*question continues on next page*)

b. Determine angle measure θ in radians. Then determine the equivalent angle measure in degrees.

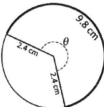

c. Determine the radius (measured in feet) of the circle with the given measures.

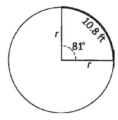

15. April is riding on a circular Ferris wheel that has a radius of 51 feet. After boarding the Ferris wheel she traveled a distance of 32.2 feet along an arc before the Ferris wheel stopped for the next rider.
    a. Make a drawing of the situation and illustrate the relevant quantities.
    b. The angle that April swept out along the arc had a measure of: i) how many degrees? ii) how many radians?

16. Consider an object moving on a circular path with a radius of 4.2 meters.
    a. How many radians are swept out when the object travels 19 meters on the circular path? Make a drawing of the situation and illustrate the quantities.
    b. How many meters does the object travel when it sweeps out 47 degrees on the circular path? Make a drawing of the situation and illustrate the quantities.
    c. If the object sweeps out $\frac{\pi}{7}$ radians on the circular path, how many meters has it traveled?
    d. Suppose the distance the object traveled on the circular path varied from 3.2 meters to 8.3 meters. How many radians did the object sweep out over this distance? Make a drawing of the situation and illustrate the quantities.

17. Using the given diagram, define a formula that relates the measures r, θ, and s. Consider r and s to be linear measures with the same units, like *number of inches*, and θ to be radians. Explain how you determined this relationship.

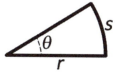

18. A common formula to calculate an angle's measure is $\theta = \frac{s}{r}$, where θ represents the measure of an angle's openness in radians, s represents the linear measure of arc length, and r represents the measure of the radius in the same units as s. Make a drawing of the situation and illustrate the quantities. Explain why the number of radius lengths in an arc length $\frac{s}{r}$ cut off by an angle results in a number of radians.

III. MODELING CIRCULAR MOTION USING THE SINE AND COSINE FUNCTION

For Problems 19 through 22, imagine a bug sitting on the end of a blasé of a fan as the blade revolves in a counter-clockwise direction. The bug is exactly 2.9 feet from the center of the fan and is at the 3 o'clock position as the blade begins to turn.

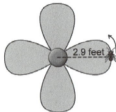

19. What quantities might you track to represent the location of a bug as it sleeps on the tip of a rotating fan blade? List as many quantities as necessary.

20. If the bug travels 6.8 feet around the circle from the 3 o'clock position, what angle measure (in radians) has been swept out by the bug?

21. a. If the bug swept out 1.23 radians on an arc as the fan blade rotated, how many feet did the bug travel along the arc?
    b. If the bug swept out 1.23 radians on an arc while sitting on a tip of a fan blade with a longer radius, would the bug travel the same number of feet as in part (a)? Explain.
    c. If a bug travels 2*r* feet (where *r* represents the length of the fan blade of various sized fans), how many radians has the bug swept out along the arc of each of these fans?

22. a. Sketch a graph of the bug's vertical distance above the horizontal diameter (measured in feet) in terms of the measure of the angle swept out by the bug's fan blade (in radians). Label the horizontal axes with "Number of radians swept out by the bug's fan blade" and begin by plotting points. Keep in mind that the vertical distance is positive when the bug is above the horizontal diameter line and negative when the bug is below it.
    b. Use your graph in part (a) to sketch a second graph that also illustrates the bug's vertical distance above the horizontal diameter in terms of the measure of the angle swept out by the bug's fan blade (in radians), but for this graph, have the output be expressed in a number of radius lengths instead of feet. What is the maximum value that the bug is away from the horizontal diameter in radius lengths? In feet? What is the relationship between 1 radius length and 2.9 feet for this situation? Why are the shapes of the graphs the same?
    c. Explain how the bug's vertical distance above the horizontal diameter changes as the bug rotates from: i) the 3 o'clock position to the 12 o'clock position; ii) the 12 o'clock position to the 9 o'clock position; iii) the 9 o'clock position to the 6 o'clock position; iv) the 6 o'clock position to the 3 o'clock position.
    d. How does the graph you constructed in part (a) change if the radius of the fan is 3.9 feet instead of 2.9 feet?
    e. Given that the graph in part (b) is represented by the function $f(\theta) = \sin(\theta)$, define a function *g* to represent the graph in part (a).
    f. For successive equal change of angle measure from the 6 o'clock position to the 3 o'clock position, how do the corresponding changes in the vertical distance change? Identify these changes on a diagram of the situation and on your graph in part (b).

For Problems 23 and 24, Ana is sitting in the bucket of a Ferris wheel. She is exactly 46.7 feet from the center and is at the 3 o'clock position as the Ferris wheel starts turning.

23. a. If Ana swept out 2.58 radians on a circular arc as the Ferris wheel rotates, how many feet did she travel? Define a function to express the number of feet Ana travels on this Ferris wheel (with a 46.7 foot radius) as a function of the number of radians θ swept out by Ana.

    b. Ana's sister is sitting on a Ferris wheel that has longer arms than Ana's does. The sister sweeps out 2.58 radians in her ride. Did the sister travel the same number of feet as Ana? If not, who traveled farther? Explain.

24. a. Sketch a graph of Ana's horizontal distance to the right of the vertical diameter (measured in feet) in terms of the measure of the angle swept out by Ana (in radians). Label the horizontal axes with "Number of radians swept out by Ana" and begin by plotting points. (The horizontal distance is positive when Ana is to the right of the vertical diameter line, and negative when Ana is to the left of the vertical diameter line.)

    b. Sketch a second graph that also illustrates how Ana's horizontal distance to the right of the vertical diameter in terms of the measure of the angle swept out by Ana in radians, but for this graph have the output be express in a number of radius lengths instead of feet. What is the maximum value that Ana is away from the vertical diameter in radius lengths and in feet? What is the relationship between 1 radius length and 46.7 feet for this situation? Why are the shapes of the graphs the same?

    c. Which of the two graphs will change if the radius of the Ferris wheel is changed? Explain.

    d. Explain how Ana's horizontal distance to the right of the vertical diameter changes as Ana rotates from: i) the 3 o'clock position to the 12 o'clock position; ii) the 12 o'clock position to the 9 o'clock position; iii) the 9 o'clock position to the 6 o'clock position; iv) the 6 o'clock position to the 3 o'clock position.

    e. Given that the graph in part (b) is represented by the function $f(\theta) = \cos(\theta)$, define a function g to represent the graph in part (a).

    f. For successive equal changes of angle measure from the 9 o'clock position to the 6 o'clock position, how do the corresponding changes in the horizontal distance change? Identify these changes on both a diagram of the situation and on your graph in part (b).

25. a. Fill in the blanks of the table for the function $f(\theta) = \sin(\theta)$.

As the input θ varies from…	the function $f(\theta)$ (increases, decreases, or remains constant)	From a value of …	to…	Resulting in a change of … in f's output
0 to $\pi/6$				
$\pi/6$ to $\pi/3$				
$\pi/3$ to $\pi/2$				
$\pi/2$ to π				
π to $3\pi/2$				
$3\pi/2$ to $5\pi/3$				
$5\pi/3$ to $11\pi/6$				
$11\pi/6$ to 2π				

    b. What patterns do you notice about how the value of $f(\theta)$ varies as θ varies from 0 to 2π?

    c. How does the change in $f(\theta)$ vary as θ varies by small, equal amounts on the interval from $\frac{\pi}{2}$ to $\frac{3\pi}{2}$?

26. a. Fill in the blanks for the table for the function $g(\theta) = \cos(\theta)$.

As the input θ varies from…	the function $g(\theta)$ (increases, decreases, or remains constant)	From a value of …	to…	Resulting in a change of … in g's output
0 to $\pi/6$				
$\pi/6$ to $\pi/3$				
$\pi/3$ to $\pi/2$				
$\pi/2$ to π				
π to $3\pi/2$				
$3\pi/2$ to $5\pi/3$				
$5\pi/3$ to $11\pi/6$				
$11\pi/6$ to 2π				

b. What patterns do you notice about how the value of $g(\theta)$ varies as θ varies from 0 to 2π?

c. How does the change in $g(\theta)$ vary as θ varies by small, equal amounts on the interval from π to 2π?

d. What conclusions can you make about the inflection points on g's graph, and the concavity of g's graph, as θ varies from π radians to 2π radians?

IV. USING THE SINE AND COSINE FUNCTIONS TO TRACK CIRCULAR MOTION

27. Determine the point (x, y) indicated on the circle.

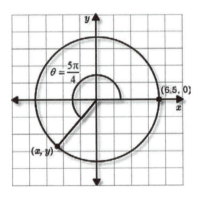

28. Determine the point (x, y) indicated on the circle.

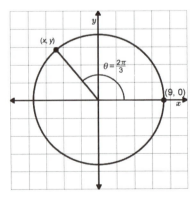

29. This figure illustrates the path of a toy racecar that begins at (4, 0) and travels 19 meters counter-clockwise on a circular path with a 4-meter radius. The racecar stops at the point (x, y).

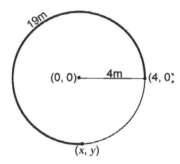

Determine the x-value (horizontal component) and the y-value (vertical component) of the racecar's position (measured in meters). Express these position coordinates in terms of the sine and cosine functions, and then determine their decimal approximations using your calculator.

30. This figure illustrates the path of a toy racecar that begins at (8, 0) and travels 38 meters counter-clockwise on a circular path with an 8-meter radius. The racecar stops at the point (x, y).

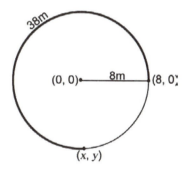

a. Determine the measure of the angle (in radians) that is formed by the ray connecting (0,0) to (8,0) and the ray connecting (0,0) to (x,y).
b. Determine the values of x and y in terms of the sine and cosine functions, and then determine their decimal approximations in meters using your calculator.
c. Define a formula that relates the horizontal component x (in radius lengths) in terms of the number of meters d the racecar has traveled along the track.
d. Define a formula that relates the vertical component y (in radius lengths) in terms of the number of meters d the racecar has traveled along the track.
e. Define a formula that relates the horizontal component x (in meters) in terms of the number of meters d the racecar has traveled along the track.
f. Define a formula that relates the vertical component y (in meters) in terms of the number of meters d the racecar has traveled along the track.

31. a. What is the general form of the ordered pair (x, y), measured in radians, of any point on a circle of radius r kilometers that forms an angle of measure θ as illustrated in the diagram?

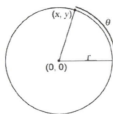

   b. Express each ordered pair (x, y) in kilometers for a circle of radius r kilometers.

32. Answer the following questions:
   a. Determine $\sin\left(\frac{\pi}{6}\right)$ and $\sin\left(\frac{5\pi}{6}\right)$. What do you notice about the two values? Explain.
   b. Determine $\sin\left(\frac{7\pi}{6}\right)$ and $\sin\left(\frac{11\pi}{6}\right)$ without using a calculator. Illustrate your answers on the unit circle.
   c. Given that $\cos\left(\frac{\pi}{6}\right) = \frac{\sqrt{3}}{2}$, determine $\cos\left(\frac{5\pi}{6}\right)$ without using a calculator. Illustrate your answer on a unit circle.
   d. Given that $\cos\left(\frac{\pi}{3}\right) = \frac{1}{2}$, determine $\cos\left(\frac{4\pi}{3}\right)$ without using a calculator. Illustrate your answer on a unit circle.

33. For parts (a)-(d), construct a unit circle and use it to determine your answers.
   a. Determine all possible values of θ that make the equation $\sin(\theta) = \frac{\sqrt{2}}{2}$ true. What is the value of $\cos(\theta)$ for each of these values of θ?
   b. Determine all possible values of θ that make the equation $\cos(\theta) = \frac{-1}{2}$ true. What is the value of $\sin(\theta)$ for each of these values of θ?
   c. Determine all possible values of θ that make the equation $\cos(\theta) = 0$ true. What is the value of $\sin(\theta)$ for each of these values of θ?
   d. Determine all possible values of θ that make the equation $\sin(\theta) = -1$ true. What is the value of $\cos(\theta)$ for each of these values of θ?

34. Construct a unit circle and label the angle measures θ and points $\left(\cos(\theta), \sin(\theta)\right)$ determined in parts (a)-(d) for values of θ between 0 and 2π (i.e., $0 < \theta < 2\pi$) that make each equations true.
   a. Given $\sin(\theta) = 0.707$ and that $\cos(\theta)$ is negative, determine the value of θ and the value of $\cos(\theta)$ and label θ and the endpoint of the arc on the unit circle you constructed.
   b. Given $\cos(\theta) = -0.99$ and that $\sin(\theta)$ is positive, determine the value of θ and the value of $\sin(\theta)$ and label θ and the endpoint of the arc on the unit circle you constructed.
   c. Given $\cos(\theta) = -0.99$ and that $\sin(\theta)$ is negative, determine the value of θ and the value of $\sin(\theta)$ and label θ and the endpoint of the arc on the unit circle you constructed.
   d. Given $\sin(\theta) = -0.91$ and that $\cos(\theta)$ is negative, determine the value of θ and the value of $\cos(\theta)$ and label θ and the endpoint of the arc on the unit circle you constructed.

V. USING THE SINE & COSINE FUNCTIONS IN APPLIED SETTINGS (TEXT: S5)

35. Suppose Jim travels 2 radians per second as he rotates on a Ferris wheel.
    a. How many radians θ does he sweep out in t seconds as the Ferris wheel rotates?
    b. Define a function v that relates Jim's *vertical distance above* the horizontal diameter of the Ferris wheel to the number of seconds t elapsed since Jim started to rotate counter-clockwise from the 3 o'clock position. (Define the relevant variables.)
    c. Define a function h that relates Jim's *horizontal distance to the right* of the vertical diameter of the Ferris wheel to the number of seconds t elapsed since Jim started to rotate counter-clockwise from the 3 o'clock position. (Define the relevant variables.)
    d. Over what time interval does Jim complete one revolution?

36. Suppose Susan travels 0.5 radians per second as she rotates on a Ferris wheel.
    a. How many radians θ does Susan sweep out in t seconds as she rotates on the Ferris wheel?
    b. Define a function j that relates Susan's *horizontal* distance to the right of the vertical diameter of the Ferris wheel to the number of seconds t elapsed since Susan started to rotate counter-clockwise from the 3 o'clock position. (Define the relevant variables.)
    c. Over what time interval does the ball complete one revolution?
    d. Sketch a graph of the function j and mark the interval of input values on which the function values complete one full cycle

37. Suppose Hope travels 0.33 radians per second as she rotates on a Ferris wheel.
    a. How many radians θ does Hope sweep out in t seconds as she rotates on the Ferris wheel?
    b. Determine a function j that defines Hope's *horizontal* distance to the right of the vertical diameter of the Ferris wheel in terms of the number of seconds t elapsed since Hope started to rotate counter-clockwise from the 3 o'clock position. (Define the relevant variables.)
    c. Over what time interval does she complete one revolution?
    d. Sketch a graph of the function j and mark the interval of input values on which the function values complete one full cycle.

38. Suppose Matthew determined that it takes 75 seconds for him to complete one revolution on a Ferris wheel.
    a. Determine the speed that Matthew travels (in radians/second) when riding the Ferris wheel.
    b. Determine a function m that defines Matthew's *vertical distance above* the horizontal diameter of the Ferris wheel in terms of the number of seconds elapsed since Matthew started to rotate counter-clockwise from the 3 o'clock position. (Define the relevant variables.)
    c. Sketch a graph of the function m and mark the interval of input values on which the function values complete one full cycle.

39. Bill extends his arm straight out holding a 2.2-foot string with a ball attached to the end of it in his hand. He twirls the ball around in a circle with his hand at the center, so that the plane in which it is twirling is perpendicular to the ground. Answer the following questions assuming the ball twirls counter-clockwise starting at the 3 o'clock position.
    a. Make a drawing of the situation and illustrate the quantities. Label the starting position of the ball.
    b. Determine a function f that defines the *vertical distance* of the ball *above* a horizontal line extending through Bill's hand (measured in feet) in terms of the measure of the angle (in radians) swept out by the ball and string as the ball twirls counter-clockwise from a 3 o'clock position. (Define variables to represent the values that the relevant varying quantities assume.)

(problem continues on next page)

   c. Suppose Bill swings the ball at a constant rate such that it makes 3 revolutions in 5 seconds. At what rate is the ball traveling in feet per second? How many radians does the ball and string sweep out per second?

   d. Write a formula that defines the distance the ball traveled *d* (measured in feet) in terms of the number of seconds *n*, since the ball began to move.

   e. Write a formula that defines the measure of the angle (in radians) swept out by the ball and string in terms of the number of seconds *n* since the ball began to move.

   f. Sketch a graph of the formula determined in part (d). On your graph, illustrate the interval of time (the values of *t*) for which the ball completes one revolution.

   g. Determine a function *g* that defines the *vertical distance* of the ball *above* a horizontal line extending through Bill's hand (measured in feet) in terms of the number of seconds since the ball began to move.

40. Caren extends her arm straight out holding a 1.3-meter string with a ball attached to the end of the string. She twirls the ball around in a circle with her hand at the center, so that the plane in which it is twirling is perpendicular to the ground. Answer the following questions assuming the ball twirls counter-clockwise starting at the 3 o'clock position.

   a. Make a drawing of the situation and illustrate the quantities. Label the starting position of the ball.

   b. Determine a function *f* that defines the *horizontal* distance of the ball *to the right* of a vertical line extending through Caren's hand (measured in meters) in terms of the measure of the angle (in radians) swept out by the ball and string as the ball twirls counter-clockwise from a 3 o'clock position. (Define variables to represent the values of the relevant varying quantities.)

   c. Suppose Caren swings the ball at a constant rate of 1 revolution per 0.7 seconds. At what rate is the ball traveling in meters per second? How many radians does the ball and string sweep out per second?

   d. Write a formula that defines the distance the ball traveled *d* (measured in meters) in terms of the number of seconds *n*, since the ball began to move.

   e. Define a formula that relates the measure of the angle (in radians) swept out by the ball and string in terms of the number of seconds *n* since the ball began to move.

   f. Determine a function *g* that defines the *horizontal* distance of the ball *to the right* of a vertical line extending through Caren's hand (measured in meters) in terms of the number of seconds since the ball began to move.

   g. Sketch a graph of the function determined in part (f). On your graph illustrate the interval of input values for which the output values complete one full cycle (i.e., the interval of *t* for which the ball completes one revolution).

   h. Suppose that the ball and string sweep out 0.25 radians per second. Write a function *j* that determines the ball's horizontal distance (measured in meters) to the right of Caren's hand in terms of the number of seconds since the ball began to move. Over what time interval did the ball complete one revolution?

VI. TRANSFORMATIONS OF THE SINE AND COSINE FUNCTIONS (TEXT: S5)

41. Given that *f* is defined by $f(\theta) = \sin(\theta)$, and *g* is defined by $g(\theta) = \sin(2\theta)$,

   a. How much does θ vary by as the argument of *g*, 2θ, varies by 2π?

   b. Determine the period of *g*.

   c. How does the graph of *g* compare with the graph of *f*?

42. Given that f is defined by $f(\theta) = \sin(\theta)$, and h is defined by $h(\theta) = \sin\left(\frac{\theta}{4}\right)$,

    a. How much does θ vary by as the argument of h, $\frac{\theta}{4}$, varies by 2π?
    b. Determine the period of h.
    c. How does the graph of h compare with the graph of f?

43. Given that f is defined by $f(\theta) = \cos(\theta)$, and j is defined by $j(\theta) = \sin\left(\frac{2\theta}{5}\right)$,

    a. How much does θ vary by as the argument of j, $\frac{2\theta}{5}$, varies by 2π?
    b. Determine the period of j.
    c. How does the graph of j compare with the graph of f?

44. a. Sketch a graph of $h(x) = \sin(5x)$, where x is in radians, on the interval $-2\pi < x < 2\pi$.
    b. What is the maximum value of h? For what value(s) of x does h assume its maximum value? (Give exact answers in terms of π.)
    c. Identify the root(s) and vertical intercept of h (give your answers as ordered pairs and express your answers in term of π).
    d. Identify the period of h on the graph.
    e. Describe how the output values of h vary as the input values vary from $\pi/2$ to π radians.
    f. What is the amplitude of h?

45. a. Sketch a graph of $g(x) = 2.5\sin\left(\frac{x}{2}\right)$, where x is in radians, on the interval $-2\pi < x < 8\pi$.
    b. What is the maximum value of g? For what value(s) of x does g assume its maximum value? (Give exact answers in terms of π.)
    c. Identify the root(s) and vertical intercept of g (give your answers as ordered pairs and express your answers in terms of π.)
    d. Identify the period of g on the graph.
    e. Describe how the output values of g vary as the input values vary from $\pi/2$ to π radians.
    f. What is the amplitude of g?

46. Let $f(\theta) = \sin(3\theta)$, where θ is in radians and is any real number. What is the period of f? How does the graph of f compare to the graph of $\sin(\theta)$? Justify your answer.

47. Let $g(\theta) = \cos\left(\frac{4\theta}{9}\right)$, where θ is in radians and is any real number. What is the period of g? How does the graph of g compare to the graph of $\sin(\theta)$? Justify your answer.

48. Let $h(\theta) = \cos(8.2\theta)$, where θ is in radians and is any real number. How does the graph of h compare to the graph of $\sin(\theta)$? Justify your answer.

49. Consider the three functions $f(x) = \cos(x)$, $g(x) = \cos(0.1x)$ and $h(x) = (3\pi x)$. All inputs are in radians.
    a. Over what interval of x will x vary by 2π?
    b. Over what interval of x will $0.1x$ vary by 2π?
    c. Over what interval of x will $3\pi x$ vary by 2π?
    d. Sketch graphs of each of the three functions on the same set of axes. Illustrate the period of each function on its graph.

50. Compare the periods of $f(x) = \sin(\frac{1}{2}x)$, $g(x) = \sin(x)$, and $h(x) = \sin(2x)$. Sketch graphs of the three functions on the same set of axes and label their periods.

51. Explain how the amplitude of *f*, defined by *f*(*x*) = 2.1cos(*x*), where *x* is in radians, compares to the amplitude of *g*, defined by *g*(*x*)=cos(*x*). Sketch graphs of *f* and *g* on the same set of axes and label the amplitude of each function.

52. Explain how the amplitude of *f*, defined by *f*(*x*) = 0.21cos(*x*), where *x* is in radians, compares to the amplitude of *g*, defined by *g*(*x*) = cos(*x*). Sketch graphs of *f* and *g* on the same set of axes and label the amplitude of each function.

53. a.   Sketch a graph of *f*(*x*) = 0.8sin(*x*), where *x* is in radians, on the interval $-3\pi < x < 3\pi$.
    b.   What is the maximum value of *f*? For what value(s) of *x* does *f* assume its maximum value?
    c.   Identify the root(s) and vertical intercept of *f* (give your answers as ordered pairs).
    d.   Identify the period of *f* on the graph.
    e.   Describe how the output values of *f* vary as the input values vary from π/2 to π radians.
    f.   What is the amplitude of *f*?

54. a.   Sketch a graph of *f*(*x*) = 5sin(*x*), where *x* is in radians, on the interval $-2\pi < x < 3\pi$.
    b.   What is the maximum value of *f*? For what value(s) of *x* does *f* assume its maximum value?
    c.   Identify the root(s) and vertical intercept of *f* (give your answers as ordered pairs).
    d.   Identify the period of *f* on the graph.
    e.   Describe how the output values of *f* vary as the input values vary from π/2 to π radians.
    f.   What is the amplitude of *f*?

55. a.   Sketch a graph of *f* given that *f*(*x*) = 5cos(*x*), where *x* is in radians, on the interval $0 < x < 2\pi$.
    b.   What is the maximum value of *f*? For what value(s) of *x* does *f* assume its maximum value?
    c.   Identify the root(s) and vertical intercept of *f* (give your answers as ordered pairs).
    d.   Identify the period of *f* on the graph.
    e.   Describe how the output values of *f* vary as the input values vary from π/2 to π radians.
    f.   What is the amplitude of *f*?

56. Given that *f* and *g* are defined as *f*(*x*) = sin(*x*) + 1.7 and *g*(*x*) = sin(*x*), sketch graphs of *f* and *g* on the same set of axis and describe how the graph of *f* compares to the graph of *g*.

57. Given that *f* and *g* are defined as *f*(*x*) = 2sin(*x*) + 1.6 and *g*(*x*) = sin(*x*), describe how the output values of *f* are related to the output values of *g*.

58. Graphs of periodic functions are provided below. Illustrate the period of each function on its graph (you may approximate the period), then determine a formula that defines the function.

a.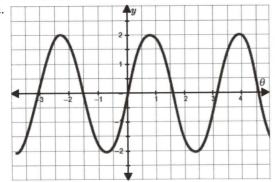

b.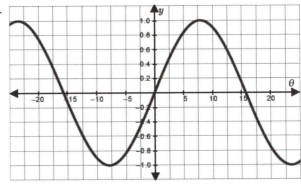

(problem continues on next page)

c.
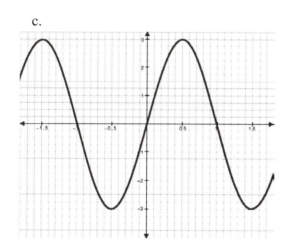

d.
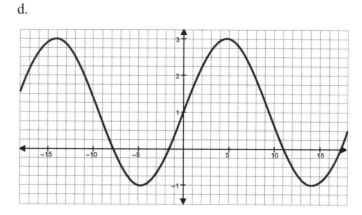

59. A picture of a water wheel (used for power in the 18th century) is illustrated below.

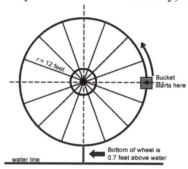

a. Sketch a graph of the *height* of the bucket above the water (in feet) as a function of the angle of rotation (measured in radians) of the bucket from the 3 o'clock position. (Define your variables and label your axes.)

b. Determine a function *f* that defines the bucket's distance *d* above the water (in feet) in terms of the angle of rotation (measured in radians) of the bucket from the 3 o'clock position. (Define your variables.)

c. Determine a function *g* that defines the bucket's distance *d* above the water (in feet) in terms of the *number of feet* the bucket has rotated counter-clockwise from the 3 o'clock position.

60. The following graph represents the average daily temperature (measured in degrees) of Athens, GA as a function of the number of days passed since June 15th. Use this graph to answer the following questions.

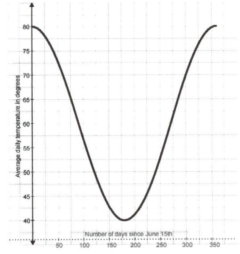

a. Choose a point on the graph and explain the meaning of this point.
b. What is the maximum average daily temperature in Athens, GA? On what day does the maximum average temperature occur?
c. What is the minimum average temperature in Athens, GA? On what day does the minimum average temperature occur?
d. How does the average daily temperature change over the time period from June 15th to December 15th? Explain your answer by discussing amounts of change in the average daily temperature for successive equal changes in the number of days passed since June 15th.
e. Explain why a trigonometric function is needed to model this phenomena. (Hint: How do the two quantities in the situation co-vary?)
f. Use the cosine function to define a function f that determines the average daily temperatures (measured in degrees) of Athens, GA when the number of days since June 15th is known..
g. Explain what each constant used in the function define in part (e) represents in the context of this situation.

61. Consider an object that is traveling on a circular path perpendicular to the ground that sweeps out a counterclockwise angle beginning from the 3 o'clock position. The angle measure (in radians) swept out by the object as a function of time (measured in seconds) is given by the function $f(t) = t^2$. Thus, the object's vertical distance (measured in radii) above the center of the circular path as a function of time (measured in seconds) is given by the function $g(t) = \sin(t^2)$.

a. Sketch a graph of the function f for $t \geq 0$. For successive equal changes of 2π in the *output* of f, what is happening to the corresponding changes of input?
b. Sketch a graph of the function g for $t \geq 0$. Explain how your findings in part (a) are reflected in the sketch of the graphs of f and g.

VII.   PERIOD, AMPLITUDE AND SHIFTS OF PERIODIC FUNCTIONS (TEXT: S5)

62. Explain why the graph of g defined by $g(x) = (x-4)^3$ is shifted 4 units to the right of the graph of f defined by $f(x) = x^3$

63. Determine how the graph of $\sin\left(\theta - \frac{\pi}{2}\right)$ compares to the graph of $\sin(\theta)$ by first setting up a table of values for both $\sin(\theta)$ and $\sin\left(\theta - \frac{\pi}{2}\right)$, then plot their graphs on the same axes. Provide an explanation to support your answer.

64. Explain why the graph of $\cos(\theta + \pi)$ is shifted π units to the left of the graph of $\cos(\theta)$.

65. Create general rules for determining how the graph of $g(\theta) = a(b\sin\theta + c\pi) + k$, where c>0 compares to the graph of $f(\theta) = \sin(\theta)$.

66. Construct the graph of h given that $h(\theta) = -2\sin\left(4\theta + \frac{\pi}{2}\right) + 3$. How does the graph of h compare to the graph of $\sin(\theta)$.

67. Construct the graph of g given that $g(\theta) = \frac{2}{3}\cos\left(\frac{\theta}{2} - \pi\right) - 5$. How does the graph of g compare to the graph of $\cos(\theta)$.

68. Explain why the output of f defined by $f(x) = \cos\left(x - \frac{\pi}{2}\right)$ is equal to the output of g defined by $g(x) = \sin(x)$ for all values of x. (Hint: Use the unit circle to justify that $\cos\left(x - \frac{\pi}{2}\right) = \sin(x)$ for all values of x.)

69. Explain how the function f defined by $f(x) = \sin\left(x + \frac{\pi}{2}\right)$ is related to the function g defined by $g(x) = \cos(x)$. Sketch graphs of f and g on the same set of axes. Identify the period and amplitude of both functions on their graphs.

70. Determine the amplitude and period of the following trigonometric functions without graphing the functions. Sketch a graph of each function after determining its period and amplitude. Label each graph's period and amplitude. *(Hint: You will need to consider the effect of some functions' values in addition to their amplitude and period).* All inputs are in radians.
    a. $f(t) = 3.2\sin(2t)$
    b. $g(x) = 0.7\cos(2x + \pi)$
    c. $h(x) = \sin(0.4x) + 1.21$
    d. $s(t) = e^2 \sin\left(\frac{2\pi}{21}t\right) + 2$
    e. $q(c) = -2.1\cos\left(\frac{\pi}{9}c + 3.2\right) - 7$
    f. $r(t) = 0.1\sin(21t - 5) + \pi$

71. Given the graphs of f and g, define a formula that defines g in terms of f.

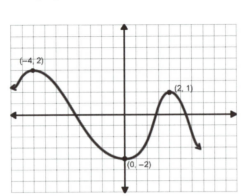

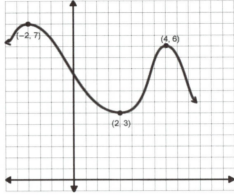

72. Graphs of periodic functions are provided below. Illustrate the period and amplitude of each function on its graph (you may approximate the period), then determine a formula that defines the function.

a.

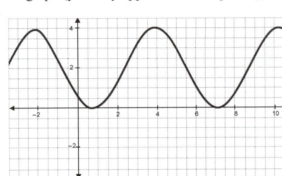

b.

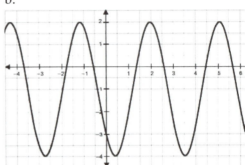

c.

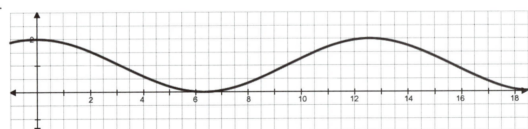

d.

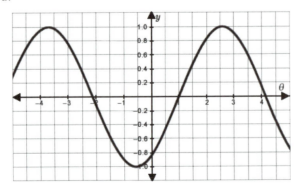

VIII.   THE TANGENT FUNCTON (TEXT: S7)

73. Evaluate each of the following then interpret the meaning of your answer.
    a. $\tan 0$

    b. $\tan \frac{11\pi}{6}$

    c. $\tan \frac{\pi}{2}$

    d. $3\tan \frac{1}{2}$

    e. $\dfrac{\sin \frac{\pi}{3}}{\cos \frac{\pi}{3}}$

    f. $\dfrac{\sin \frac{5\pi}{4}}{\cos \frac{5\pi}{4}}$

74. Draw an angle with a measure of θ that meets the set of requirements given.
    a. $\tan(\theta)$ is positive and $\cos(\theta)$ is negative

    b. $\sin(\theta)$ is positive and $\tan(\theta)$ is negative

    c. $\cos(\theta)$ is positive and $\tan(\theta)$ is negative

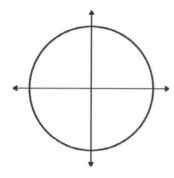
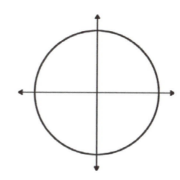
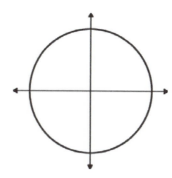

75. What values of θ make the equation $\tan(\theta)=0$ true when $0<\theta< \pi$?

76. Let $f(\theta)=\tan(\theta)$. What is the input variable for the inverse relation of the function f and what does it represent? What is the output variable for the inverse relation of the function f and what does it represent?

77. The graph of $f(\theta)=\tan(\theta)$ is given. Use your understanding of function translations to make a rough sketch of a graph for each of the following functions.

    a. $g(\theta)=\tan(2\theta)$

    b. $h(\theta)=\tan \frac{\theta}{3}$

    c. $j(\theta)=2\tan(\theta)$

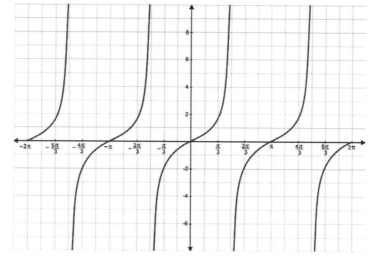

Since we interpret the tangent function as returning the slope of the terminal ray for an angle measure θ (measured in radians), we can also define the identity, $\tan(\theta) = \frac{y}{x}$ where (x, y) is the intersection point of the terminal ray and the circle. Use this information for Problems 78 and 79.

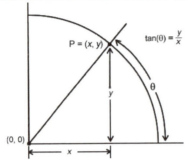

78. Explain why the identity $\tan(\theta) = \frac{y}{x}$ must be true.

79. As θ varies from 0 to $\frac{\pi}{2}$,
    a. how does x vary?
    b. how does y vary?
    c. how does the ratio $\frac{y}{x}$ vary?
    d. Explain how the variation of x and y on the interval from $\theta = 0$ to $\theta = 1$ impacts the variation of $\tan(\theta) = \frac{y}{x}$ on this same interval of θ.

IX. INVERSE TRIGONOMETRIC FUNCTIONS (TEXT: S6)

80. Given that $\sin\left(\frac{\pi}{6}\right) = \frac{1}{2}$, determine the value of $\arcsin\left(\frac{1}{2}\right)$. Justify your answer.

81. If $\cos\left(\frac{\pi}{3}\right) = \frac{1}{2}$, determine the value of $\cos^{-1}\left(\frac{1}{2}\right)$. Justify your answer.

82. If $\tan\left(\frac{\pi}{4}\right) = 1$, determine the value of $\tan^{-1}(1)$. Justify your answer.

83. If $\tan\left(\frac{2\pi}{3}\right) = -\sqrt{3}$, determine the value of $\tan^{-1}\left(-\sqrt{3}\right)$. Justify your answer.

84. If $\cos\left(\frac{5\pi}{4}\right) = -\frac{\sqrt{2}}{2}$, determine the value of $\cos^{-1}\left(-\frac{\sqrt{2}}{2}\right)$. Justify your answer.

85. Explain why $\sin^{-1}\left(\sin(\theta)\right) = \theta$ when $\frac{-\pi}{2} \le \theta \le \frac{\pi}{2}$.

86. For what values of θ is the statement $\tan^{-1}\left(\tan(\theta)\right) = \theta$ true? Explain your reasoning.

87. Explain why $\cos\left(\cos^{-1}\left(\frac{3\theta}{2}\right)\right) = \frac{3\theta}{2}$. when $\frac{-2}{3} \le \theta \le \frac{2}{3}$.

88. Demonstrate on a circle how one could estimate the value of $\cos^{-1}(-0.8)$.

89. Evaluate $\sin^{-1}(0.7)$. What does your answer represent?

90. Evaluate the following expressions in radians.
    a. arccos (0.5)   b. 3 arcsin (−1)   c. 5 arccos (0)

91. Evaluate each of the following for $x = 0.5$.
    a. $\sin\left(x^{-1}\right)$ b. $(\sin(x))^{-1}$ c. $\sin^{-1}(x)$
    d. Explain the meaning conveyed by the notation in each of (a-c) above.

92. a. Solve the following equations for values of θ, given that $\frac{-\pi}{2} < \theta < \frac{\pi}{2}$
      i. $\tan(\theta) = 1$ ii. $\tan(\theta) = -1$ iii. $3\tan(2\theta) = -8$ iv. $2\tan(\theta) = -\frac{1}{3}$
    b. In part (a), why was it necessary to restrict the domain to $\frac{-\pi}{2} < \theta < \frac{\pi}{2}$ in order to provide a single answer in parts (i) through (iv)?

93. Solve the following equations for θ if possible. If not, say why not.
    a. $3\sin(\theta) = 2$ b. $\cos(\theta) = -0.3$ c. $\frac{1}{2}\sin(\theta) = -0.2$

94. Solve each equation for θ, given that $0 < \theta < 2\pi$. Be sure to show your work. Check your answer by graphing and labeling your solutions on your graphs.
    a. $\sin(4\theta) = 0.5$ b. $2\cos(\theta) = -\frac{1}{3}$ c. $\sin(\theta) = 0$ d. $8\cos(2\theta) = -6$

95. Determine the measure of the angle θ (in radians) indicated on the circle.

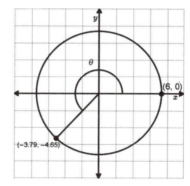

96. Determine the measure of the angle θ (in radians) indicated on the circle.

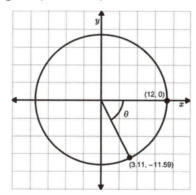

97. Given the following points on a circle of radius 4 meters centered at the origin, determine the corresponding counter-clockwise arc length (in the same units as the coordinates of each point) from the 3 o'clock position (the ray that connects (0,0) and (4,0)).

    a. (−0.713, 3.936) meters
    b. (−0.924, 0.382) radii
    c. (−3.282, −2.286) meters
    d. (−0.924, −0.382) radii

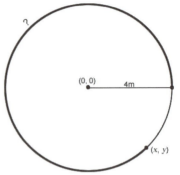

98. Given the following circle and undetermined angle measures of α and θ radians, answer the following questions.

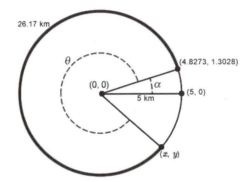

    a. What is the value of θ, the measure of the angle indicated in the figure above?

    b. What quantity does the value $\frac{1.3028}{5}$ represent? What is the unit of measure for this quantity?

    c. What quantity does the value $\frac{4.8273}{5}$ measure? What is the unit of measure for this quantity?

    d. What is the value of α, the measure of the angle indicated in the figure above?

99. An arctic village maintains a circular cross-country ski trail that has a radius of 2.5 kilometers. A skier started skiing from position (−1.76777, −1.76777), measured in kilometers, and skied counter-clockwise for 3.927 kilometers, where he paused for a brief rest. (Consider the circle to be centered at the origin). Determine the ordered pair (in both kilometers and radii) on the coordinate axes that identifies the location where the skier rested.

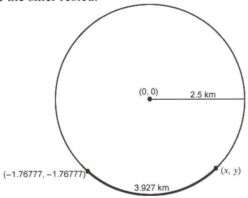

© 2015 Carlson, Oehrtman, and Moore

100. A skier started skiing from position (2.4136, 0.6513), measured in kilometers, and skied counter-clockwise for 13.09 kilometers where he paused for a brief rest. Determine the ordered pair (in both kilometers and radii) on the coordinate axes that identifies the location where the skier rested.

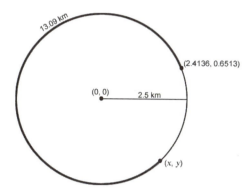

101. Porter is sitting in a bucket of a Ferris wheel at the 3 o'clock position. The Ferris wheel has a radius of 45 feet. The Ferris wheel begins moving counterclockwise.

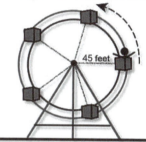

   a. Define a function to represent Porter's vertical distance (in feet) above the horizontal diameter of the Ferris wheel in terms of his angle of rotation (in radians) as he rotates counter clockwise from the 3 o'clock position

   b. By what angle(s) has Porter rotated from the 3 o'clock position when he is 27 feet above the horizontal diameter of the Ferris wheel? (Set up an equation and show how to use the arcsine or $\sin^{-1}$ function to determine one of the angles.)

   c. Define a function to represent Porter's horizontal distance to the right of the vertical diameter of the Ferris wheel in terms of his amount of counter clockwise rotation (in radians) from the 3 o'clock position.

   d. What angle(s) has Porter rotated when he is −5 feet to the right of the vertical diameter of the Ferris wheel? ((Set up an equation and show how to use $\cos^{-1}$ to solve for one of the angles.)

102. A biologist tracked the deer population in a rural area of Wisconsin and found that the deer population in this area was cyclic. He used his data to find a function to approximate the population over a year-long period. The function g defined by $g(t) = 1125 - 875\cos\left(\frac{\pi}{6}t\right)$, $0 < t < 12$, represents the number of deer $g(t)$ in terms of the number of months t since November 1, 2009. Use the arccosine function to determine the month(s) in which the deer population was at least 600.

103. Wind turbines, or windmills, are currently used in an attempt to produce green energy.

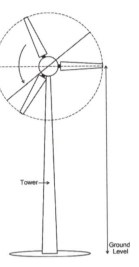

   a. On the diagram to the right, choose values for the height of the turbine's tower (measured in meters) and for the turbine's radius (in a number of meters not equal to 1). Label these values on the diagram.
   b. Sketch a graph of the distance of the fan blade's tip (measured in meters) above the horizontal diameter of the windmill as a function of measure of the angle (in radians) swept out by the fan blade from a 3 o'clock starting position.
   c. Define a function *f* that will create the graph you sketched in part (b). Define variables to represent the values the quantities assume (include units).
   d. Sketch a graph of the distance of the fan blade's tip (measured in meters) *above the ground* as a function of the measure of the angle (in radians) swept out by the fan blade.
   e. Define a function *g* that will create the graph you sketched in part (d). Define variables to represent the values the quantities assume (include units). Explain how the function's definition reflects the fact that you are measuring distances *above the ground*.

104. Complete the following tasks given that the wind turbine in the previous problem rotates at a rate of 3 radians every 15 seconds.
   a. Define a function *h* that relates the measure of the angle (in radians) swept out by the fan blade as a function of time elapsed since the fan blade started rotating from the 3 o'clock position. Sketch the graph of *h*. On your graph, identify the interval of time needed such that the fan completes one revolution.
   b. Sketch a graph of the distance of the fan blade's tip (in meters) above the horizontal diameter of the windmill as a function of the number of seconds that have elapsed since the fan started rotating from the 3 o'clock position. On your graph, identify the interval of time needed such that the fan completes one revolution.
   c. Define a function *f* that will generate the graph you sketched in part (b). Define variables to represent the values the quantities assume (include units). Explain how the function conveys the rate (in radians per second) at which the fan blade rotates.

105. Monica and Louie go to the fair and decide to take a ride on the Ferris wheel. The Ferris wheel operator tells them that if they can keep track of how far they have traveled while in their seat, they will win a prize. The catch is that they have to be able to tell him how far they have traveled any time he asks for it. The only information he gives them is that the radius of the Ferris wheel is 35 feet. After agreeing to his challenge, Monica notes there are 12 total carts equally spaced around the Ferris wheel. Louie and Monica are also the first group to board at the bottom of the Ferris wheel.
   a. Make a drawing of the situation and illustrate the quantities. In the diagram, label the location of Louie and Monica when boarding the Ferris wheel as well as the location of Louie and Monica when the last cart is filled.
   b. When the cart after their cart is being filled, what is the length of the arc (measured in feet) traveled by Louie and Monica? What is the length of the arc (measured in radii) traveled by Louie and Monica? Label these measurements on your diagram above, explicitly identifying the arc length being measured.
   c. After the operator has filled up four total carts, including Louie and Monica's, he asks them for the distance they have traveled. What is the length of the arc (measured in feet) traveled by Louie and Monica? What is the length of the arc (measured in radii) traveled by Louie and Monica? Label these measurements on your diagram, explicitly identifying the arc length being measured.

(problem continues on next page)

d. The Ferris wheel starts again and the operator continues to fill carts. Finally, as the last cart is being filled, the operator asks Louie and Monica how far they have traveled. What is the length of the arc (measured in feet) traveled by Louie and Monica? What is the length of the arc (measured in radii) traveled by Louie and Monica? Label these measurements on your diagram above, explicitly identifying the arc length being measured.

106. Consider the Ferris wheel situation described in the previous problem.
   a. What was Louie and Monica's vertical distance *above the center of the Ferris wheel* (measured in feet) when they first boarded the Ferris wheel? What was Louie and Monica's vertical distance *above the center of the Ferris wheel* (measured in radii) when they first boarded the Ferris wheel? (Consider their vertical distance to be negative when below the center of the Ferris wheel and positive when above the center of the Ferris wheel.)
   b. What is Louie and Monica's vertical distance *above the center of the Ferris wheel* (measured in feet) when the cart *after* Louie and Monica's cart is being loaded? What is Louie and Monica's vertical distance *above the center of the Ferris wheel* (measured in radii) when the cart *after* Louie and Monica's cart is being loaded?
   c. Define a function *f* that relates Louie and Monica's vertical distance above the center of the Ferris wheel (measured in radii) as a function of the angle they swept out (measured in radians) during their ride.
   d. Define a function *g* that relates Louie and Monica's vertical distance above the center of the Ferris wheel (measured in feet) as a function of the angle they swept out (measured in radians) during their ride.
   e. If the bottom of the Ferris wheel is 11 feet above the ground, define a function *h* that relates Louie and Monica's distance above the ground (measured in feet) as a function of the measure of the angle (in radians) they swept out during their ride.
   f. Sketch a graph of the function *h* defined in part (e). Label two points on the graph and explain the meaning of these points in the context of the problem.
   g. Suppose the radius of the Ferris wheel were doubled. Would this change the function from part (e)? If so, re-write the function for this new Ferris wheel. Explain your reasoning.
   h. Suppose the radius of the Ferris wheel were doubled. Would this change the function from part (c)? If so, re-write the function for this new Ferris wheel. Explain your reasoning.

107. Carlos is the last person to board a Ferris wheel that has a radius of 52 feet. The counterclockwise arc swept out by Carlos is measured from where he got on, *at the bottom of the Ferris wheel*. Since Carlos is the last one to board, the Ferris wheel starts up and doesn't stop again until the ride is over.
   a. Draw a diagram that show's Carlos's location and the angle swept out by Carlos at an arbitrary moment in time.
   b. Define a function *f* that relates Carlos's *vertical* distance above the center of the Ferris wheel (measured in feet) as a function of the measure of the angle (in radians) swept out by Carlos. Over what interval of input will Carlos complete one revolution?
   c. Define a function *g* that relates Carlos's *horizontal* distance to the right of the center of the Ferris wheel (measured in feet) as a function of the measure of the angle (in radians) swept out by Carlos. Over what interval of input will Carlos complete one revolution?
   d. Suppose Carlos's cart is traveling 0.3 radii per second on a circular path as the wheel rotates. Define a function *h* that relates Carlos's *horizontal* distance to the right of the center of the Ferris wheel (measured in feet) as a function of the time since beginning the ride (measured in seconds). How long will it take for Carlos to complete one revolution?
   e. Marilyn claimed that she saw Carlos's cart travel at a speed of 3.5 radii per second. Is that possible? Explain. (Hint: Determine the cart's speed in miles per hour.

(problem continues on next page)

f. Suppose that Carlos's cart traveled at 0.4 radii per second. Define a function *j* that relates Carlos's *horizontal* distance to the right of the center of the Ferris wheel (measured in feet) as a function of the time since beginning the ride (measured in seconds). How long will it take for Carlos to complete one revolution?

g. Suppose Carlos's cart is traveling 0.2 radii per second. Define a function *k* that relates Carlos's *horizontal* distance to the right of the center of the Ferris wheel (measured in feet) as a function of the time since beginning the ride (measured in seconds). How long will it take for Carlos to complete one revolution?

h. Sketch a graph of each function defined in parts (d)-(g). Label your axes. Identify the interval of input over which Carlos completes one revolution.

108. Lucia boards a Ferris wheel with a radius of 42.5 feet. When Lucia boards, the Ferris wheel is at the bottom of its rotation. The angle swept out as Lucia rotates is measured from the 6 o'clock position.
a. Make a drawing of the situation and illustrate the quantities.
b. Graph Lucia's *vertical distance above* the ground (measured in feet) as a function of the measure of the angle (in radians) swept out by Lucia from the bottom of the Ferris wheel. (Define your variables and label your axes.)
c. Define a function *f* that relates Lucia's vertical distance above the ground (measured in feet) as a function of the measure of the angle (in radians) swept out by Lucia from the bottom of the Ferris wheel. (Define variables to represent the values the quantities assume (include units).
d. Suppose that the Ferris wheel travels at a constant rate of 7.5 feet per second and makes no stops. Express the distance traveled by Lucia (in feet) as a function of the number of seconds since the Ferris wheel began to turn.
e. Define a function *g* that relates Lucia's *vertical distance above* the ground (measured in feet) as a function of the number of seconds elapsed since the Ferris wheel began to turn. How many seconds will it take for Lucia to complete one revolution?
f. Sketch a graph of the function you defined in part (e). On your graph, identify the interval of input values on which the function values complete one full cycle (i.e., the interval of *t* for which Lucia completes one revolution)

109. A buoy sitting in the ocean bobs up and down such that it moves vertically 4.2 feet between its high and low points every 9 seconds. When you peered out the window of your cabin the buoy was at its highest point and dropped to its lowest point (i.e., sea level) before returning back to its highest point, completing the full cycle in 9 seconds.
a. What quantities in this situation are changing? Define variables to represent the values the quantities assume (include units). Create a graph to represent the covariation of these two quantities. (The graph should illustrate how the values of the two quantities change together.) Be sure to label your axes.
b. Define a function *f* that relates the vertical position of the buoy (measured in feet) in terms of the number of seconds that have elapsed since you began watching the buoy. Be sure to explicitly identify the input and output quantities of the situation (e.g., say more than the output is distance, say *what distance it is*, such as the vertical distance of the buoy above its halfway position between the high and low points).
c. Describe what each variable and constant value in your function represents.
d. Sketch a graph of your function from part (b) and describe the co-variation of the input and output quantities over a time period of 5 seconds to 6.1 seconds
e. Use the function you graphed to determine the times when the buoy reaches the halfway position between the high and low points.

110. The London Eye is a large Ferris wheel that is a famous landmark of London. The function below models a person's height above the ground (in feet) as a function of the number of minutes they've been on the Eye.

$$f(t) = -221\cos(\tfrac{\pi}{15}t) + 221$$

   a. What is the amplitude of f and what does this value represent in the context of the problem>
   b. What is the period of f and what does this value represent in the context of the problem? Explain how you determined the period.
   c. Define a function g that relates the height of a person off the ground (measured in feet) as a function of the distance traveled (measured in feet) using the cosine function.
   d. Alter the function f to reflect the situation in which the London Eye rotates twice as fast.
   e. Alter the function f to reflect the situation in which the radius of the London Eye is doubled.
   f. Sketch graphs of the given function and the functions you defined in parts (e) and (f) on the same set of axes.

1. Recall that the measure of an angle corresponds to the fraction of a circle's circumference that the angle subtends.

   a. By using the hypotenuse of the triangle below as a radius, construct a circle with its center at the vertex of the angle such that the radius of the circle is 5. Then, determine $\sin(\theta)$ without determining the value θ. Hint: Consider the process of determining the value θ and consider how the output of the sine function relates to this value.

   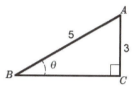

   b. Using the hypotenuse of the triangle below as a radius, construct a circle with its center at the vertex of the angle such that the radius of the circle is 5.43. Then, determine $\sin(\theta)$ without determining the value θ.

   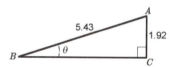

2. a. Using the hypotenuse of the triangle below as a radius, construct a circle with its center at the vertex of the angle such that the radius of the circle is 4. Find the length of the segment BC and then determine $\cos(\theta)$ without determining the value θ.

   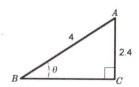

b. Use the diagram below to determine $\cos(\theta)$ and $\sin(\theta)$. Explain your reasoning.

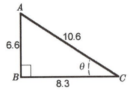

3. a. Using the hypotenuse of the triangle below as a radius, construct a circle with its center at the vertex of the angle such that the radius of the circle is 11.88. Then, determine $\tan(\theta)$ without determining the value θ.

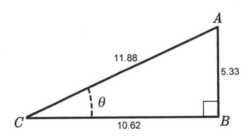

4. Using the hypotenuse of the triangle below as a radius, construct a circle with its center at the vertex of the angle such that the radius of the circle is 3.96.

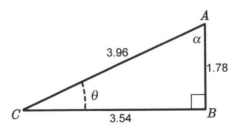

a. How could you use your knowledge of the sine and/or cosine functions to determine the measure of θ? Explain your reasoning.

b. Determine the measure of α.

5. Use the diagram below to determine each value.

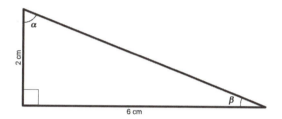

a. Determine $\sin \alpha$.

b. Determine $\cos \alpha$.

c. Determine $\tan \alpha$.

d. Determine $\sin \beta$.

e. Determine $\cos \beta$.

f. Determine $\tan \beta$.

g. Determine the measures of angles α and β.

6. a. Given that $\sin(\theta) = -\dfrac{3}{4}$ and that θ is in quadrant III, determine $\cos(\theta)$ and $\tan(\theta)$.

b. Given that $\tan(\alpha) = -\dfrac{2}{3}$ and that α is in quadrant II, determine $\sin(\theta)$ and $\cos(\theta)$.

1. Consider a person standing 12 feet from a light pole. The individual notices that the light casts a shadow of his body that is 5 feet long. If the individual is 5 feet, 6 inches tall, how tall is the light pole?
    a. Create a diagram and label the known and unknown quantities.

    b. Determine the height of the light pole and justify your solution.

    c. Determine an alternative method to find the height of the light pole. (e.g., If you used (inverse) trigonometric functions to solve part (b), use similar triangles to determine the unknown height. If you used similar triangles to solve part (b), use (inverse) trigonometric functions to determine the unknown height.)

2. A plane leaves a local air force base and travels due east. A radar station 45 miles south of the base tracks the plane and determines that the angle formed by the base, the radar station, and the plane is initially changing by 1.6 degrees per minute.
    a. Sketch a diagram of the situation and label the known and unknown quantities.

    b. Define a function that relates the plane's distance from the radar station *d* (measured in miles) as a function of the elapsed time *t* (measured in minutes) since leaving the base.

    c. Define a function that relates the plane's distance from the air force base *c* (measured in miles) as a function of the elapsed time *t* since leaving the base (measured in minutes).

d. Use your function in part (c) to complete the following table.

Number of minutes t since leaving the base	The plane's distance c (measured in miles) from the base	$\dfrac{\Delta c}{\Delta t}$
0		
0.5		
1		
1.5		
2		

Describe how the plane's distance from the air force base c (measured in miles) changes as a function of the elapsed time during the first 2 minutes.

e. Use your function in part (c) to complete the following table.

Number of minutes t since leaving the base	The plane's distance c (measured in miles) from the base	$\dfrac{\Delta c}{\Delta t}$ (miles per minutes)
42		
45.5		
49		
52.5		
56		

f. Describe how the plane's distance from the air force base c changes as a function of the elapsed time between 42 and 56 minutes since leaving the base. Are these values realistic? Why or why not?

g. Graph the relationship between the plane's distance from the air force base and the amount of time since leaving the base. What does the graph and the tables in parts (d) and (e) convey about the speed of the plane?

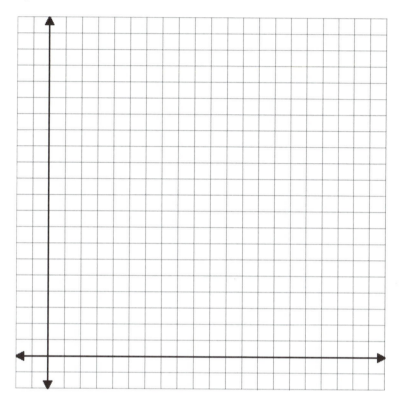

© 2015 Carlson, Oehrtman, and Moore

1. Answer parts (a)-(d) using the following right triangle.

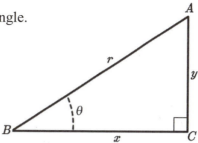

a. Use the Pythagorean Theorem to determine the relationship between x, y, and r.

b. Determine a formula relating $\cos(\theta)$ and x.

c. Determine a formula relating $\sin(\theta)$ and y.

d. Rewrite the relationship determined in part (a) using only $\cos(\theta)$, $\sin(\theta)$, and r.

2. Simplify the following expressions to create a trigonometric identity. In other words, rewrite the following expressions in their most simplified form.

a. $\dfrac{\sin x}{\csc x} + \dfrac{\cos x}{\sec x} = ?$

b. $\dfrac{\tan(x)\cot(x)}{\csc(x)} = ?$

c. $\sin(x)\tan(x) + \cos(x) = ?$

$\sin \frac{o}{H}$ $\cos \frac{A}{H}$ $T \frac{o}{A}$

~~cot~~ $\sec \frac{H}{A}$

$\tan^2 = \sec^2 - 1$

3. Algebraically prove the following identities. In other words, choose the expression on one side of the equals sign and simplify this expression to equal the expression on the other side of the equals sign.

a. $1 + \tan^2(x) = \sec^2(x)$

$1 + \frac{\sin^2 x}{\cos^2 x} = \frac{1}{\cos^2 x}$

$\frac{\cos^2 x + \sin^2 x}{\cos^2 x} = \frac{1}{\cos^2 x}$

$\frac{1}{\cos^2 x} = \frac{1}{\cos^2 x}$

b. $\dfrac{1 - \cot(x)}{\cos(x)} = \sec(x) - \csc(x)$

$= \frac{1}{\cos x} - \frac{1}{\sin x}$

$\frac{\sin x - \cos x}{\cos x} = \frac{\sin x - \cos x}{(\sin x)(\cos x)}$

$\frac{\sin}{\cos x} \cdot \frac{1}{\sin x (\cos x)} - \frac{\cos}{\sin x} \cdot \frac{1}{\sin x (\cos x)}$

$\frac{1}{\sin x (\cos x)} - \frac{1}{\sin x (\cos x)}$

c. $(\sec(x) - \tan(x))^2 = \dfrac{1 - \sin(x)}{1 + \sin(x)}$

$\sec^2 x - \tan^2 x$

$+1 \quad = \frac{1 - \sin x}{1 + \sin x}$

$\sec^2 x + \tan^2 x - 2\sec x \tan x$

$\sec^2 x + \tan^2 x - 2 \frac{1}{\cos x} \cdot \frac{\sin}{\cos x}$

$\sec^2 x - 2\sec x (\tan x) + \tan^2 x = \frac{1 - \sin(x)}{1 + \sin(x)}$

$\frac{1}{\cos^2 x} - 2\sec x (\tan x) + \frac{\sin^2 x}{\cos^2 x}$

$\frac{1}{\cos^2 x} - 2\frac{\sin x}{\cos^2 x} + \frac{\sin^2 x}{\cos^2 x}$

$\frac{1 - 2\sin x + \sin^2 x}{\cos^2 x} = \left(\frac{1 - \sin x}{\cos^2 x}\right)^2$

$\frac{\cos^2 x}{\cos^2 x}$

$\frac{1 - \sin^2 x}{(1 - \sin^2 x)}$

4. Use the trigonometric identity $\cos(x + y) = \cos(x)\cos(y) - \sin(x)\sin(y)$ to determine the exact value of $\cos\left(\dfrac{\pi}{12}\right)\cos\left(\dfrac{2\pi}{3}\right) - \sin\left(\dfrac{\pi}{12}\right)\sin\left(\dfrac{2\pi}{3}\right)$.

© 2015 Carlson, Oehrtman, and Moore

1. Surveyors were interested in determining the approximate size of the lake shown below, so they set up their instruments in three different locations, and measured the angles created. They were also able to directly measure the distance between Surveyor A and Surveyor B as shown.

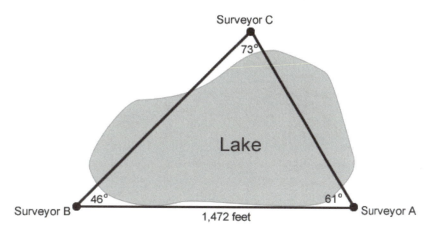

a. The triangle created is not a right triangle, however we can apply the techniques of right triangle trigonometry by forming right triangles out of the given triangle. Draw a diagram below to show a few different ways that this is possible.

b. The lines you drew in part (a) to create right triangles are called *altitudes* (lines drawn from a vertex of a triangle perpendicular to the opposite side). Which altitude(s) will create a diagram that allows us to apply right triangle trigonometry techniques to solve for the lengths of the remaining sides of the triangle? Defend your choice.

c. Find the distance from Surveyor A to Surveyor C and from Surveyor B to Surveyor C.

2. Two ships have each determined their distance from a lighthouse on the shore. An observer at the lighthouse determines the angle between the ships. Determine the distance between the ships.

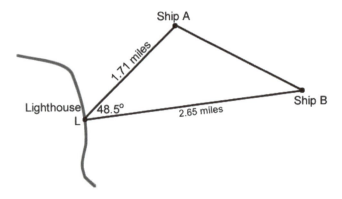

3. Find the length of the missing sides and the missing angle measures in the following triangles.
   a.

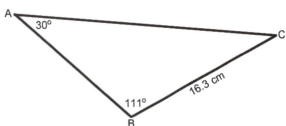

   b.

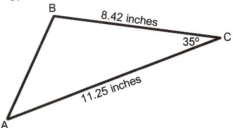

4. Find the area of the following triangle.

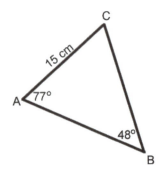

1. Consider the non-right triangle *ABC* shown below (with side lengths *a*, *b*, and *c*), with altitude *BD* (whose length is *h*).

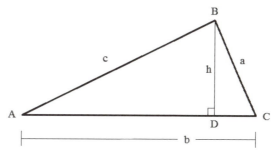

   a. Write an expression that calculates the value of *h* if $m\angle A$ and *c* are known.

   b. Write an expression that calculates the value of *h* if $m\angle C$ and *a* are known.

   c. What do the results of parts (a) and (b) tell us?

   d. Now, suppose we draw the altitude *CE* (whose length is *k*).

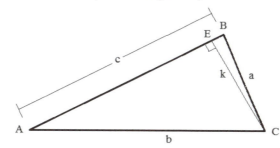

   i. Write an expression that calculates the value of *k* if $m\angle A$ and *b* are known.

   ii. Write an expression that calculates the value of *k* if $m\angle B$ and *a* are known.

   iii. What do the results of parts (i) and (ii) tell us?

Since the triangle in Problem #1 was a generic triangle (no specific side lengths or angle measures were given), our conclusions will generalize to *any* triangle. This generalization produces the rule known as *The Law of Sines*. It is summarized below.

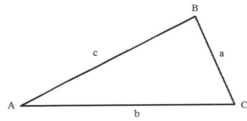

For any non-right triangle *ABC*, we have that $\dfrac{a}{\sin(m\angle A)} = \dfrac{b}{\sin(m\angle B)} = \dfrac{c}{\sin(m\angle C)}$.

2. Use the Law of Sines to find the value of *x* in each of the following triangles.

a.

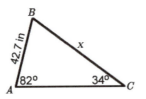

b.

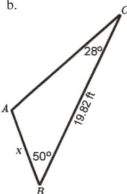

3. Use the Law of Sines to find the values of *x* and *y* in each of the following triangles.

a.

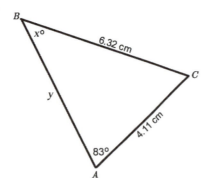

b.

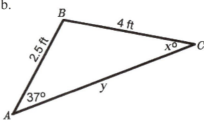

4. In many cases the Law of Sines works perfectly well and returns the correct missing values in a non-right triangle. However, in some cases the Law of Sines returns *two* possible measurements. This is the case in the triangle below.

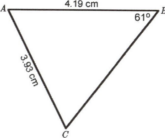

We begin the solution process, shown below.

$$\frac{4.19}{\sin(m\angle C)} = \frac{3.93}{\sin(61°)}$$

$$\frac{\sin(m\angle C)}{4.19} = \frac{\sin(61°)}{3.93}$$

$$\sin(m\angle C) = \frac{\sin(61°)}{3.93} \cdot 4.19$$

$$\sin(m\angle C) = 0.9325$$

$$m\angle C = \sin^{-1}(0.9325)$$

It's at this point that we get two possible answers for $m\angle C = \sin^{-1}(0.9325)$.

a. Why does $m\angle C = \sin^{-1}(0.9325)$ return two possible answers for $m\angle C$ in this triangle?

b. If we assume that the drawing is more or less to scale, which value of $m\angle C$ makes more sense? Use this value of $m\angle C$ to determine the remaining measurements of the triangle.

c. Determine the dimensions of the second possible triangle with the given measurements, then draw the triangle approximately to scale.

5. There are also two possible triangles with the given measurements shown below. Find the measurements of both triangles and draw each triangle approximately to scale.

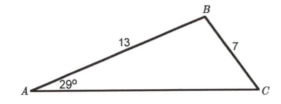

1. The Law of Sines is an extremely useful tool for finding missing lengths or missing angle measures in non-right triangles. However, there are times when there isn't enough information to apply the Law of Sines. Examine the following non-right triangles and explain why the Law of Sines cannot be used to find missing lengths or missing angle measures.

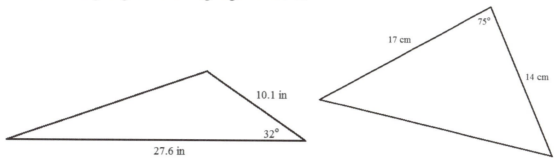

Let's take a situation like the triangles shown in Problem #1: a triangle in which the measure of one angle is known and the length of the included sides are known. For example, suppose we know $m\angle C$ and a and b in the following triangle and we are interested in finding the length c.

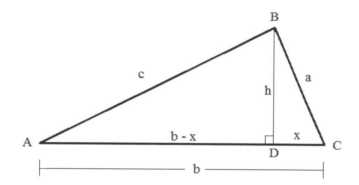

First, observe that the Pythagorean theorem assures that $h^2 + (b-x)^2 = c^2$ and $x^2 + h^2 = a^2$ (so $h^2 = a^2 - x^2$). Also, we have that $\cos(m\angle C) = \dfrac{x}{a}$, so $x = a \cdot \cos(m\angle C)$. Then

$$h^2 + (b-x)^2 = c^2$$
$$h^2 + (b-x)(b-x) = c^2$$
$$h^2 + b^2 - 2bx + x^2 = c^2$$
$$(a^2 - x^2) + b^2 - 2b + x^2 = c^2 \qquad \bullet \text{ since } h^2 = a^2 - x^2$$
$$a^2 - \cancel{x^2} + b^2 - 2bx + \cancel{x^2} = c^2$$
$$a^2 + b^2 - 2bx = c^2$$
$$a^2 + b^2 - 2b(a \cdot \cos(m\angle C)) = c^2 \qquad \bullet \text{ since } x = a \cdot \cos(m\angle C)$$
$$a^2 + b^2 - 2ab \cdot \cos(m\angle C) = c^2$$

The conclusion $a^2 + b^2 - 2ab \cdot \cos(m\angle C) = c^2$ is called *The Law of Cosines*. Depending on the known information, equivalent forms of this law include

$$a^2 = b^2 + c^2 - 2bc \cdot \cos(m\angle A)$$

$$b^2 = a^2 + c^2 - 2ac \cdot \cos(m\angle B)$$

$$c^2 = a^2 + b^2 - 2ab \cdot \cos(m\angle C)$$

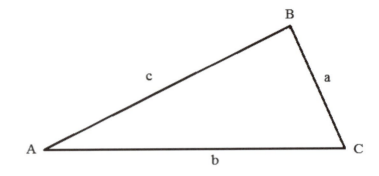

2. Use the Law of Cosines to find the value of x in each of the following triangles.

a.

b.

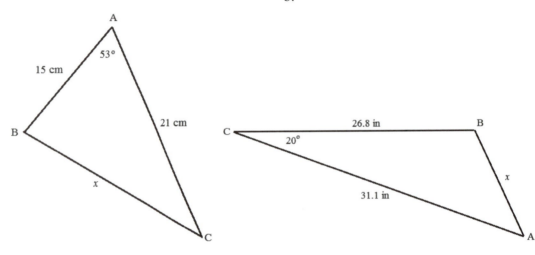

3. Use the Law of Cosines to find the value of *x* in each of the following triangles.

a.

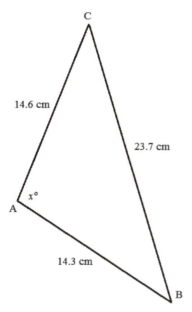

b.

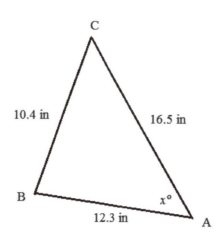

4. The Law of Cosines and the Law of Sines can be applied to other shapes as well by first breaking them into triangles. The diagram below shows how this is done.

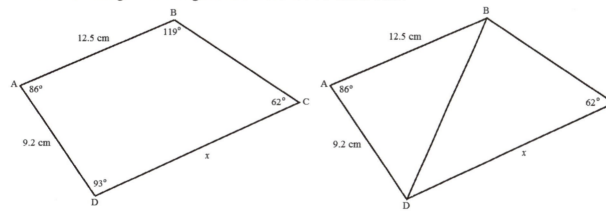

Use the given information to find the value of *x*.

5. Find the value of x in the following diagram.

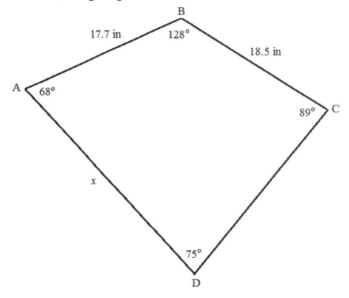

I. RIGHT TRIANGLE TRIGONOMETRY

1. Given the circle below, centered at the origin and point (x, y) in the first quadrant, state the meaning of $\sin(\theta)$, $\cos(\theta)$, and $\tan(\theta)$ in terms of x and y; then in terms of o, h, and a.

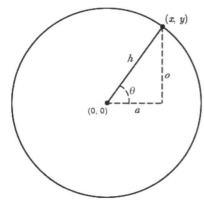

2. Determine the measure of the angles (in both degrees and radians) and the length of each side for each right triangle below.

a.

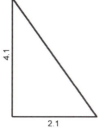

b.

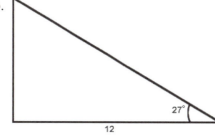

c.

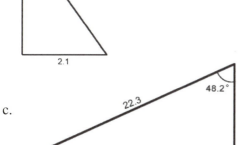

d.

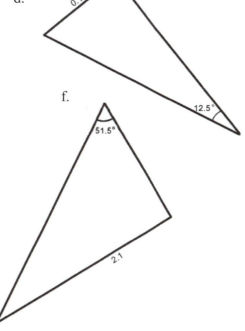

e.

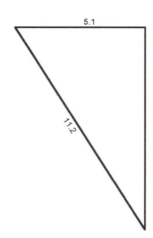

f.

___

3. Use the diagram below to answer the following questions.

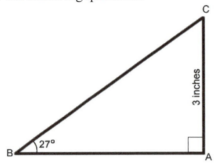

    a. Find the length of hypotenuse $\overline{BC}$.

    b. Find the length of side $\overline{AB}$.

    c. Find $m\angle C$.

4. Use the diagram below to answer the following questions

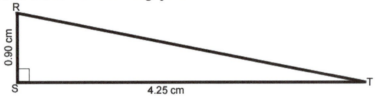

    a. Find the length of the hypotenuse $\overline{RT}$.

    b. Find $m\angle R$.

    c. Find $m\angle T$

5. For the following circles fill in the missing value(s). Assume θ is measured in radians.

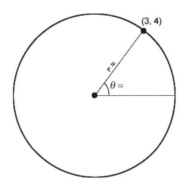   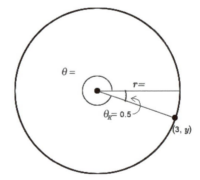

    (see two more circles on the following page)

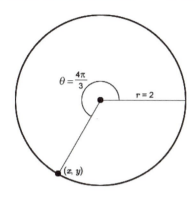

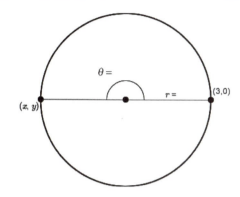

6. a. Given that $\cos(\theta) = \dfrac{1}{3}$ and that θ is in quadrant I, determine $\sin(\theta)$ and $\tan(\theta)$.

    b. Given that $\tan(\alpha) = -\dfrac{2}{5}$ and that α is in quadrant IV, determine $\sin(\theta)$ and $\cos(\theta)$.

II. Right Triangle Trigonometry Applications

7. Consider the following situation when responding to (a)-(h): A 10-foot ladder is leaning against a wall with the bottom of the ladder positioned 2 feet from the wall. A painter comes along and pulls the ladder away from the wall at a constant rate of 0.75 feet per second.

    a. Fill in the values of the following table.

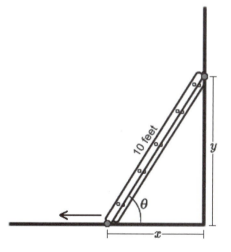

t	x	y	θ *(degrees)*	$\dfrac{\Delta y}{\Delta t}$	$\dfrac{\Delta \theta}{\Delta t}$
0					
1					
2					
3					
4					
5					
6					
7					
8					
9					
10					
10.67					

b. Describe how the distance of the bottom of the ladder from the wall *x* changes with respect to the elapsed time *t* measured in seconds.
c. Describe how the height of the top of the ladder from the ground *y* changes with respect to the elapsed time *t* measured in seconds.
d. Describe how the angle of the ladder from the ground *θ* changes with respect to the elapsed time *t* measured in seconds.
e. Define a function *f* that relates the vertical distance, *y*, as a function of the horizontal distance, *x*.
f. Define a function *g* that relates the angle measure *θ* as a function of the horizontal distance *x*.
g. Define a function *h* that relates the horizontal distance, *x*, as a function of the elapsed time *t*.
h. Define a function *j* that relates the angle measure, *θ*, as a function of the elapsed time *t*.

8. After Jack planted his magic beans, his neighbor Jill watched the beanstalk grow. When the top of the beanstalk was at her eye level (5 feet), Jill began tracking the growth of the beanstalk from a stationary position 27 feet from the base of the beanstalk. After 125 seconds, she noted that the beanstalk reached the first cloud and estimated that her line of site was at an 85° angle with respect to the ground.
a. Draw a diagram of the situation and label each known and unknown quantity.
b. How tall was the beanstalk 125 seconds after Jill began tracking the growth of the beanstalk?
c. How fast did the beanstalk grow, given that the beanstalk grew vertically at a constant rate of change of distance with respect to time?
d. Define a function that relates the height of the beanstalk from the ground *h* as a function of the angle of Jill's line of sight with respect to the ground as she watched the top of the stalk grow.

9. The stadium light poles for a soccer field are located 20 feet from the sideline. The lamp at the top of each 52 foot tall pole needs to be set at an angle upward from the pole so that the center of the light beam hits the soccer field 36 feet inside the sideline toward the center of the field.
a. Draw a diagram of the situation and label each known and unknown quantity.
b. At what angle should the lamp be set to meet the stated conditions?
c. Based on the angle found in part (b), if that angle increases by 1°, what is the increase in the distance from the base of the pole to where the center of the light hits the field?

10. A platform is to be built on a tall pole. To keep the platform from swaying in the wind, a total of eight guy-wires will be attached to the pole. The pole has a height of 166 feet. Four guy-wires will be attached to the pole at a height of 140 feet. These wires are 161 feet in length. The other four guy-wires will be attached to the pole at a height of 109 feet. These guy-wires are 131 feet in length.
a. Draw a diagram of the situation and label each known and unknown quantity.
b. At what distance from the base of the pole will the longer guy-wires be fastened to the ground?
c. At what distance from the base of the pole will the shorter guy-wires be fastened to the ground?
d. What angle will the longer guy-wires make with the ground? What angle will the shorter guy-wires make with the ground?

11. A shoreline observation post is located on a cliff such that the observer is 310 feet above sea level. When initially spotted, the angle of depression (the angle at which an observer needs to look down) from the observation post to an approaching ship was 4.8°.
a. Draw a diagram of the situation and label each known and unknown quantity.
b. How far out to sea was the ship when it was first spotted?
c. After watching the ship for 47 seconds, the angle of depression was 5.9°. Assuming the ship was moving at a constant rate, estimate the velocity of the ship.
d. If the ship were to continue on its course at the same velocity, how much time from the second reading will elapse before the ship crashes into the shore?

12. An architect designs a building that has an overhang above a south-facing floor-to-ceiling window. This window is 10 feet high. At the latitude of this building, the sunlight hits the ground at a 79.5° angle on the summer solstice (June 21) and 32.5° on the winter solstice (December 21).
    a. Label the following diagram with the known quantities.

    b. How long should the overhang project from the side of the building so that the sunlight hits just at the bottom of the window on the summer solstice?
    c. How long should the overhang project from the side of the building so that the sunlight hits just at the bottom of the window on the equinox (March 21 and October 21), when the sunlight's angle is 56° with respect to the ground?
    d. If the overhang is built to shade the window from spring equinox to fall equinox, as described in part (c), how high up the window will the sun hit on the winter solstice?

13. An airport radio tower, which is 80 meters tall, is tracking a plane that is flying towards the airport, which is located at sea level. The plane is maintaining an altitude of 9.5 kilometers and flying at a speed of 482 kilometers per hour. Initially, the plane is 177 kilometers (the line of sight distance) from the radio tower.
    a. Draw a diagram of the situation and label each known and unknown quantity.
    b. What is the initial angle of elevation from the top of the radio tower to the plane? What is the initial horizontal distance of the plane to the radio tower?
    c. Define a function that relates the plane's angle of elevation from the top of the tower as a function of the time elapsed since beginning to track the plane.
    d. When is the plane directly above the tower?

14. While driving on a straight highway across the flat plains of eastern Wyoming, you notice a snow-capped mountain peak in the distance, directly ahead. When stopping at a gas station for coffee and snacks you use your electronic angle-measuring tool and find that the angle of elevation to the top of the mountain was 1.2637°. Later, after stopping at a rest area 25.0 miles further along the road, you find that the angle of elevation from there was 2.015°.
    a. Draw a diagram of the situation and label each known and unknown quantity.
    b. From the information above, determine the height of the mountain in feet.
    c. How far are you from the *top* of the mountain at each location (the line of sight distance)?
    d. You look to the south and notice a familiar mountain. On the map you find that you are 36 miles from the base of this mountain. You also know the mountain is 0.34 miles tall. What angle should the measuring tool return?

III. TRIGONOMETRIC IDENTITIES

15. Answer the following questions:
    a. Explain in your own words what the identity $\sin(2x) = 2\sin(x)\cos(x)$ conveys. Say more than "they are equal" (i.e., what does it mean to say that they are equal?).
    b. Given that $\sin(x) = 2/3$ and $\cos(x) > 0$, determine the values of $\cos(x)$, $\tan(x)$, $\csc(x)$, $\sec(x)$, and $\cot(x)$ without using inverse trigonometric functions to determine the value of x.
    c. Simplify the following expression until it is expressed as a single trigonometric function:
    $$\frac{\sin(\theta)}{1 + \cos(\theta)} + \cot(\theta)$$
    d. Given the following right triangle, determine the values in terms of x of the six trigonometric functions with an input of θ radians.

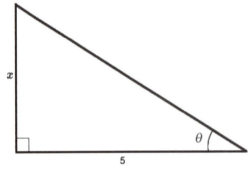

    e. Use the sum and difference identities to prove $\cos(x - \pi/2) = \sin(x)$ and $\sin(x + \pi/2) = \cos(x)$ for all values of x.

16. For each of the following, complete the identity by simplifying. In other words, rewrite each of the following in its most simplified form.
    a. $\dfrac{\tan x \cdot \cot x}{\csc x} = ?$
    b. $\sec x - \sec x \cdot \sin^2 x = ?$

17. Using the following geometric proof, fill in the missing lengths (A and B) and explain how each length is determined.
    $$\sin (x + y) = \sin (x) \cos (y) + \cos (x) \sin (y)$$

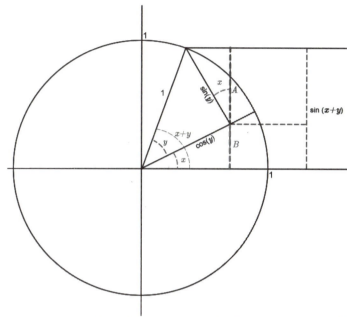

18. A 14 ft. ladder is leaning against a wall, such that its base is 12.9 ft from the wall. Then the base of the ladder is moved toward the wall into a new position, such that the angle the base makes with the floor has doubled from the angle of its initial position.
    a. Draw a diagram that represents the situation.
    b. Use the identity $\cos(2x) = 2\cos^2(x) - 1$ to determine the distance of the base of the ladder from the wall in its new position.
    c. Find the distance of the base of the ladder from the wall in its new position using another method.

19. Use the sum and difference identities and the Pythagorean identities to algebraically prove the following double angle identities.
    a. $\sin(2x) = 2\sin(x)\cos(x)$ c. $\cos(2x) = 2\cos^2(x) - 1$
    b. $\cos(2x) = \cos^2(x) - \sin^2(x)$ d. $\cos(2x) = 1 - 2\sin^2(x)$

20. Use the sum and difference identities to algebraically prove the following sum to product identities.
    a. $\dfrac{1}{2}(\sin(x - y) + \sin(x + y)) = \sin(x)\cos(y)$

    b. $\dfrac{1}{2}(\cos(x - y) + \cos(x + y)) = \cos(x)\cos(y)$

    c. $\dfrac{1}{2}(\cos(x - y) - \cos(x + y)) = \sin(x)\sin(y)$

21. Algebraically prove the following identities.
    a. $\tan(\theta) + \dfrac{\cos(\theta)}{1 + \sin(\theta)} = \sec(\theta)$ d. $\sec(\theta) = \dfrac{1}{\sqrt{1 - \sin^2(\theta)}}$
    b. $\tan^2(\theta) = \sec^2(\theta) - 1$ e. $\dfrac{1 - \cos(2\theta)}{2} = \sin^2(\theta)$
    c. $\dfrac{\tan(\theta)}{\sqrt{1 + \tan^2(\theta)}} = \sin(\theta)$ f. $\dfrac{1 + \cos(2\theta)}{2} = \cos^2(\theta)$

Pythagorean Identities:	*Sum/Difference Identities:*

$$\sin^2(\theta) + \cos^2(\theta) = 1$$
$$\tan^2(\theta) + 1 = \sec^2(\theta)$$
$$1 + \cot^2(\theta) = \csc^2(\theta)$$

Sum/Difference Identities:

$$\sin(\theta_1 + \theta_2) = \sin(\theta_1)\cos(\theta_2) + \cos(\theta_1)\sin(\theta_2)$$
$$\sin(\theta_1 - \theta_2) = \sin(\theta_1)\cos(\theta_2) - \cos(\theta_1)\sin(\theta_2)$$
$$\cos(\theta_1 + \theta_2) = \cos(\theta_1)\cos(\theta_2) - \sin(\theta_1)\sin(\theta_2)$$
$$\cos(\theta_1 - \theta_2) = \cos(\theta_1)\cos(\theta_2) + \sin(\theta_1)\sin(\theta_2)$$

Double Angle Identities:

$$\sin(2\theta) = 2\sin(\theta)\cos(\theta)$$
$$\cos(2\theta) = \cos^2(\theta) - \sin^2(\theta)$$
$$\cos(2\theta) = 2\cos^2(\theta) - 1$$
$$\cos(2\theta) = 1 - 2\sin^2(\theta)$$

Product to Sum Identities:

$$\sin(\theta_1)\cos(\theta_2) = \frac{1}{2}(\sin(\theta_1 - \theta_2) + \sin(\theta_1 + \theta_2))$$
$$\cos(\theta_1)\cos(\theta_2) = \frac{1}{2}(\cos(\theta_1 - \theta_2) + \cos(\theta_1 + \theta_2))$$
$$\sin(\theta_1)\sin(\theta_2) = \frac{1}{2}(\cos(\theta_1 - \theta_2) - \cos(\theta_1 + \theta_2))$$

Half Angle Identities:

$$\sin^2\left(\frac{\theta}{2}\right) = \frac{1 - \cos(\theta)}{2}$$
$$\cos^2\left(\frac{\theta}{2}\right) = \frac{1 + \cos(\theta)}{2}$$

IV. APPLYING TRIGONOMETRY TO NON-RIGHT TRIANGLES

22. The diagram below shows some measurements taken for a new bridge. How long is the bridge?

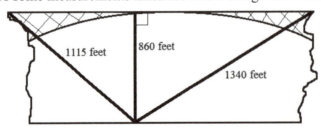

1115 feet   860 feet   1340 feet

23. Three surveyors have placed themselves at three locations around the edge of a canyon, measuring the angles between them. Surveyor A and Surveyor C are 1,359.2 feet apart. What is the distance between Surveyor A and Surveyor B? Surveyor B and Surveyor C?

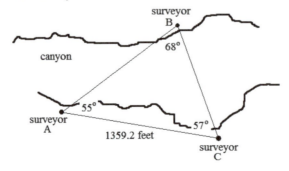

24. Standing on top of a building, you use a laser measuring tape to measure your distance from the top and bottom of a nearby statue. How tall is the statue?

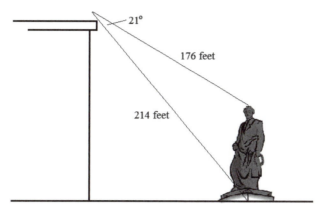

25. Find the value of x in each of the following triangles.

a.

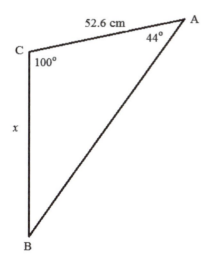

b.

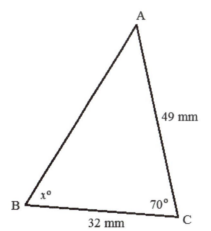

c.

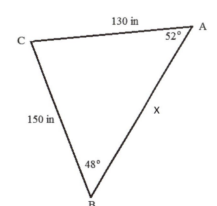

d.

e.

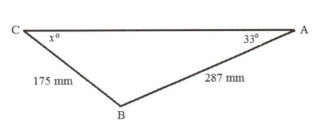

f.

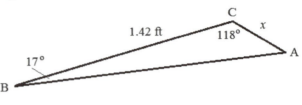

26. Find the area of the following triangles.

a.

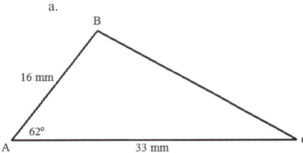

b.

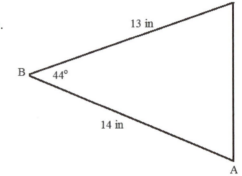

V. THE LAW OF SINES

The Law of Sines

$$\frac{a}{\sin(m\angle A)} = \frac{b}{\sin(m\angle B)} = \frac{c}{\sin(m\angle C)}$$

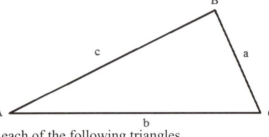

27. Use the Law of Sines to find the value of x in each of the following triangles.

a.

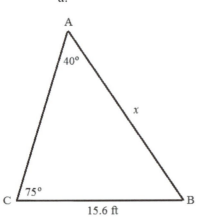

b.

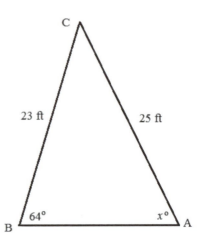

c.

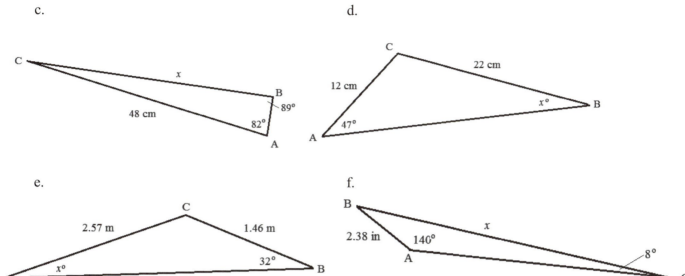

d.

e.

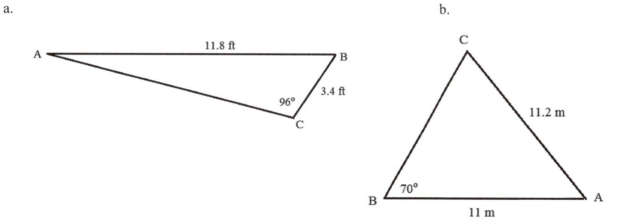

f.

28. Find the length of the missing sides and the missing angle measures in the following triangles.

a.

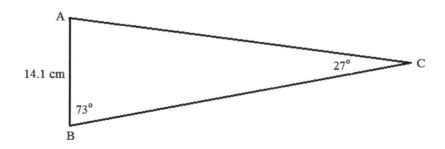

b.

c.

29. There are two possible triangles with the given measurements shown below. Find the measurements of both triangles and draw each triangle approximately to scale.

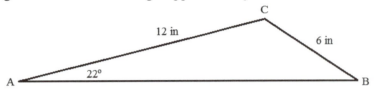

30. There are two possible triangles with the given measurements shown below. Find the measurements of both triangles and draw each triangle approximately to scale.

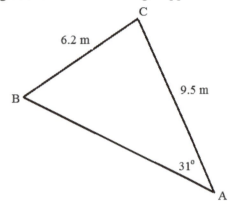

31. There are two possible triangles with the given measurements shown below. Find the measurements of both triangles and draw each triangle approximately to scale.

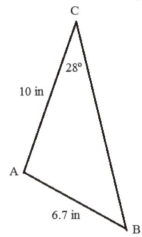

VI. THE LAW OF COSINES

The Law of Cosines

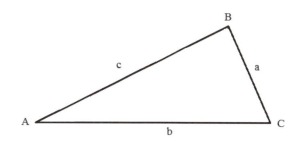

$$a^2 = b^2 + c^2 - 2bc \cdot \cos(m\angle A)$$

$$b^2 = a^2 + c^2 - 2ac \cdot \cos(m\angle B)$$

$$c^2 = a^2 + b^2 - 2ab \cdot \cos(m\angle C)$$

32. Use the Law of Cosines to find the value of x in each of the following diagrams.

a.

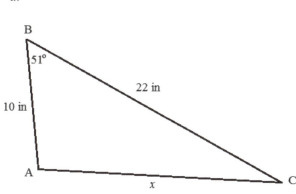

b.

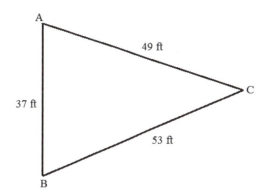

c.

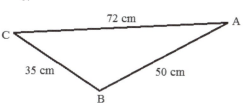

d.

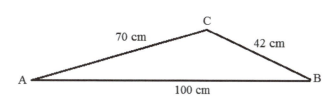

e.

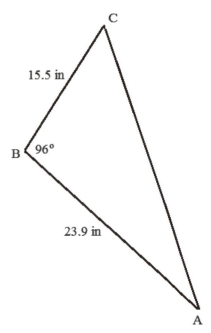

f.

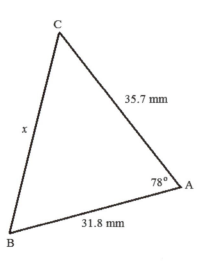

33. Find the lengths of the missing sides and the measures of the missing angles.

a.

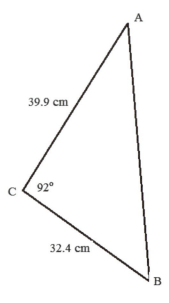

b.

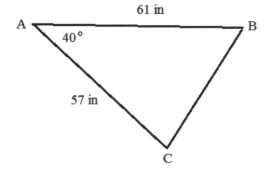

c.

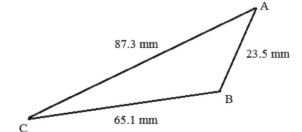

34. Find the value of x in each of the following diagrams.

a.

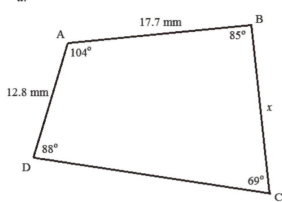

b.

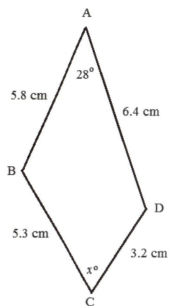

c.

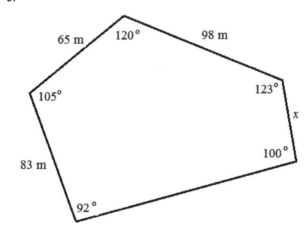